8/09

FLIGHTLESS
BIRDS

Recent Titles in
Greenwood Guides to the Animal World

Nocturnal Animals
Clive Roots

Hibernation
Clive Roots

FLIGHTLESS BIRDS

■ Clive Roots

Greenwood Guides to the Animal World

GREENWOOD PRESS
Westport, Connecticut • London

Library of Congress Cataloging-in-Publication Data

Roots, Clive, 1935–
   Flightless birds / Clive Roots.
      p. cm.—(Greenwood guides to the animal world, ISSN 1559–5617)
   Includes bibliographical references and index.
   ISBN 0–313–33545–1 (alk. paper)
   1. Flightless birds. I. Title. II. Series.
   QL677.79.F55R66   2006
   598.16—dc22   2006010619

British Library Cataloguing in Publication Data is available.

Library of Congress Catalog Card Number: 2006010619
ISBN: 0–313–33545–1
ISSN: 1559–5617

First published in 2006

Greenwood Press, 88 Post Road West, Westport, CT 06881
An imprint of Greenwood Publishing Group, Inc.
www.greenwood.com

Printed in the United States of America

The paper used in this book complies with the
Permanent Paper Standard issued by the National
Information Standards Organization (Z39.48–1984).

10 9 8 7 6 5 4 3 2 1

Contents

Preface

To have wings and not use them is an anomaly, but during their evolution many birds lost the power of flight. When food was readily available, predators were absent, and the climate reasonable year-round, they had no reason to fly, and their muscles deteriorated through lack of use until eventually they were grounded permanently. But evolution is a process of improvement, and just like the numerous flying birds known only from fossils, many flightless species also perished as the circumstances which favored flightlessness changed; they were at a disadvantage and could not compete.

About two thousand years ago the world supported some very diverse flightless birds. They included Madagascar's huge elephant birds, which were much larger than ostriches; the stocky moas and tiny wrens, which had evolved in New Zealand; waterbirds living on high Andean lakes; and large penguin-like birds, which thrived in the cold northern Atlantic. Most had been flightless for many years, an indication that their flightless lifestyle was a successful one. Others, mainly more recent converts to flightlessness, had been blown by gales to remote oceanic islands where they also found paradise, and because they lacked pressure to become airborne, the result was predictable. But like the natural extinctions that occurred throughout the ages, these modern-day flightless birds were also affected by progress: not the improvements of evolutionary change, but exposure to human progress such as hunting, the sudden loss of habitat, and the many alien predators against which they were defenseless. Dozens of flightless species were exterminated and the future of many others is now in doubt.

These unique birds, which decided long ago not to fly, now suffer the consequences of their actions in a world changed by mankind. Their lifestyles and their fate are the subjects of this book.

Introduction

Flightlessness is mainly a condition of the juvenile bird, grounded from hatching until the growth of its primary or flight feathers provides the necessary airlift. These feathers are a bird's largest and can take several weeks or many months to grow, depending on the species. Tiny birds like the hummingbirds can fly when they are just three weeks old, but chicks of the wandering albatross, which has the largest wingspan of all birds when adult, are not fully feathered until the age of nine months. There are exceptions to this general rule, like the chicks of the mound-builders or megapodes, which have wing feathers when they hatch (because of their extra-long incubation and development period) and can fly almost immediately. Also, the young of some ground-dwelling game birds, such as the pheasants and partridges, can fly when they are just a few days old, which improves their ability to avoid predators. These chicks are said to be precocial, which is the bird equivalent of precocious, but means the same—very advanced for their age. To be grounded for life is a much rarer condition in birds, but it occurs naturally in a small percentage of all living species. It also happens unnaturally: accidentally as a side effect to birds that have been continually controlled for many generations by humans, and deliberately to certain zoo and park species.

The ability to fly actually depends more on the quality of a bird's wing muscles than upon the size of its wings. The major muscles needed for flapping wings are on a bird's breast and like all muscles must be used regularly to maintain their size and tone or they quickly deteriorate. The unnatural loss of flight has occurred in several domesticated birds like the chicken and turkey, which have very inactive lives compared to their wild ancestors. They rarely need to fly, and in many cases their pens are too small anyway, especially the wire cages known as laying batteries, in which they can barely open their wings, and have consequently lost their flight due to inactivity. Lack of flight, through lack of opportunity or desire, resulted in their wing muscles eventually degenerating to the point where they can hardly lift the bird off the ground. They cannot sustain the powered or flapping flight needed to gain height and distance, which is unnecessary anyway for the domesticated bird, because it does not have to escape from its enemies, search for food, or migrate. In addition, these birds are now bred for a specific purpose, usually just for their meat or eggs, and when supermarkets needed larger and cheaper chicken breasts the commercial poultry industry complied by genetically selecting the most suitable breeding stock to produce these qualities.

Although domesticated chickens and turkeys have huge breasts of white muscle meat, it is not the right kind of muscle for flapping wings and getting airborne. This kind of flight loss was therefore a side effect of providing larger chicken breasts and huge Thanksgiving turkeys, which are double the weight of their wild ancestors still living in the Oregon woods, where the record tom weighed 37 pounds (16.8 kg).

Domesticated Turkey *Many long-domesticated birds, especially the selectively bred commercial breeds such as the chicken and turkey, are flightless due to the deterioration of their wings through lack of use, and their great increase in size. This grounded domesticated turkey is almost double the weight of its flying ancestors.*
Photo: Clive Roots

The deliberate and unnatural control of flight also occurred at the hands of bird keepers, where it has been used as a tool of captive wild bird management for many years in zoos, bird gardens, and parks, but only in large species, especially waterfowl (ducks, geese, and swans) and the cranes. The permanent loss of flight allows these birds to be kept loose on lakes and in ornamental gardens, and this grounding is achieved through making a bird lopsided when it attempts to fly. This unevenness or asymmetry of the wings—making one shorter than the other—is achieved in several ways, but the most common one is the practice of pinioning.

This is a minor surgical procedure which involves the removal of the last segment of one wing, equivalent to the fingers of a human hand, from which grow the large primary or flight feathers. This makes a bird's "wing-loading"—the area of wing in relation to its body weight—uneven, and as a result it cannot lift off. As many people find this rather disconcerting, certainly disfiguring and even morally offensive, the practice is now being replaced by tendon surgery, which leaves the bird's wings intact, but prevents them from opening fully.

However, this book is not concerned with the inability of young birds to fly before the growth of their wing feathers, nor with the results of our animal husbandry activities. It is about the natural and permanent flightlessness of adult wild birds whose ancestors could fly long ago, and the reasons for their loss of flight. It is also about those birds in which flightlessness is a temporary condition, and the many others whose flight has been compromised, although they are not yet totally grounded.

Of all the animal attributes which have most impressed modern humans, such as the bloodhound's sense of smell, the eagle's eyesight, and the dolphin's aquatic grace and speed, the ability to fly is undoubtedly the characteristic they would most like to emulate. Mankind has defied gravity and soared high above the earth in a variety of powered craft and powerless gliders, but self-powered flight—to emulate the birds and fly under one's own power—has long been an obsession. Even in the late nineteenth century, the principles of aerodynamics were not fully understood and there were ludicrous attempts at flapping flight. Men jumped off roofs with huge feathered wings strapped to their arms and shoulders, which they attempted to flap quickly before their inevitable crash landing. Beating the air in this way to imitate birds is an impossible task for us anyway, as our pectoral muscles are inadequate for powering the great expanse of wings needed to support such a large body weight in the air's thin medium. A wingspan of at least 20 feet (6 m) would likely be needed to gain lift-off, and our pectoral muscles could neither lift nor flap such wings. Birds know well the limits of lifting a body into the air by wing power, and the great bustard (*Otis tarda*) is the world's heaviest flying bird, with adult males reaching a weight of 46 pounds (21 kg) and getting airborne with some difficulty on wings spanning only 7 feet (2.1 m). The wild turkey (*Meleagris gallopavo*) is also a very large flying land bird, with several specimens exceeding 33 pounds (15 kg) and one a record 37 pounds (16.8 kg) shot recently in Oregon. Several birds have grown heavier, nine living species in fact, comprising seven kinds of ratites (flightless birds like the ostrich, emu, and rhea) and the two largest penguins, but they grew large only after losing their flight. Also, birds have several advantages over humans when it comes to flying, including lightweight bones and wings that almost touch as they rotate backwards and upwards in readiness for the downstroke, whereas we can only pull our arms back just over 180 degrees at right angles to our shoulders. With arms, muscles, and joints inadequate to the task, self-powered flight by humans is possible only through leg power, peddling like fury to turn the propellor of an ultralight machine.

Flying gave birds many advantages over the terrestrial mammals and reptiles. They could quickly escape a potential threat, they could nest in places inaccessible to most predators, and they could fly long distances in search of mates, food, water, or a better climate. Flying appears to be such a great boon, it seems inconceivable that a bird should lose the ability, especially when the advantages of flightlessness are quite small in comparison. One advantage is the capacity to grow very large, as the restrictions that flight places on size no longer apply. Increased size was an advantage in a cold environment as this decreased the ratio of a bird's surface area to its volume, thus conserving heat, which was a definite asset for the great auk of the North Atlantic Ocean and the emperor penguin and king penguin of Antarctica. With increased size there is also the potential for increased speed, which was clearly an advantage for some terrestrial or land-dwelling birds, plus the development of powerful legs and feet, which gave them a good chance of either outrunning their predators or kicking them to death. However, only the ratites have such leg power and several flightless birds have not increased in size. Morever, many land birds are not only good runners but take to the air as a last resort, which seems a far more sensible arrangement than losing the ability to fly and having to rely solely on running. There were certainly savings in energy for birds that became flightless, especially the considerable energy no longer required for flapping their wings or for regrowing their large wing feathers annually after each moult, which is a major drain on a bird's resources. Yet all other living birds, over 8,000 species of them in fact, do not seem unduly stressed by flying or the necessity to provide the large amounts of almost pure protein required to replace their wing feathers at least once annually after moulting.

The lack of large feathered wings, which could become waterlogged and heavy, would seem to be an advantage for aquatic birds by improving their diving and swimming capability, but this has not always been the case. Certainly the flightless penguin's wings, reduced to long, paddle-like flippers, are perfect for underwater swimming, but the flying auks, rather stumpy marine birds like the puffin and razorbill, manage very well propelling themselves underwater with their feathered wings, using the same short, fast wing-beats that they use when airborne. Other waterbirds such as the cormorants and grebes have evolved into very capable swimmers using their feet alone, and wings capable of flight have not affected their ability to chase and catch their prey underwater. The cormorant's wing feathers are not even waterproof, and after swimming underwater it perches on a branch or rock and holds its wings out to dry.

Originally there were few, if any, disadvantages to becoming flightless, otherwise birds would not have lost their flying ability. Unforseeable natural environmental disasters and evolutionary changes were another matter, and major climate change and the rise of the mammalian carnivores took their toll of birds that were unable to migrate or escape by flight. However, the principal disadvantage of being flightless only became apparent within the last two millennia in the form of the havoc wreaked by humans, including the loss of habit, overhunting, and the introduction of alien animals to oceanic islands, to which the flightless species had no answer.

If even the natural disadvantages of being permanently grounded outweighed the advantages, why would birds become flightless in the first place? Simply because it was inevitable. Birds did not consciously choose to lose their powers of flight; it just happened when they did not use their wings for the very purpose they had evolved.

Ostrich *Flightless birds require other means of locomotion, and in many land species, especially the ratites or cursorial birds, flightlessness was accompanied by a great increase in size and strong legs and feet to support the extra weight. In Africa this powerful body and great speed (the fastest bird on land) enabled the ostrich to survive alongside the many evolving mammalian predators, especially the lion and hyena.*

Photo: Courtesy Harcourt Index

When their behavior and the environment did not require them to fly, they used their wings less frequently, then hardly at all, and eventually lost their use altogether. It was the original case of use them or lose them. During a bird's embryonic stages simple genetic changes produced this flightlessness in the adult bird within a few generations when they were placed in a situation where flight was unnecessary and they did not use their wings. Flightlessness is simply the inability to flap the wings and gain and maintain altitude. Dropping off a branch and coming straight down in a flurry of wings is not flight, nor is clambering up a tree to glide down to a lower point. Several flightless birds actually do this, as do a number of other animals including the flying squirrels, flying lizards, and even flying frogs, but they cannot fly.

Flightless birds are mostly island birds. In fact, with the exception of the ostrich, emu, rheas, and a subspecies of the cassowary, the flightless land birds all live on islands, generally smaller ones not exceeding 35,000 square miles (135,000 square km). The flightless waterbirds all occur on sea islands and along coastlines, with the exception of three freshwater species that live on remote mountain lakes in South America. The living flightless birds can be placed in four major categories: the ratites, penguins, rails, and the waterbirds, which leaves just two unusual species—the kakapo (*Strigops habroptilus*) and kagu (*Rhinochetus jubatus*). There are thirty-eight species and endemic island subspecies or races of flightless land birds and twenty-six flightless waterbirds, which is only a small percentage of the 8,700 known species of living birds. The flightless land birds have either lost their wings, like the cassowary and kiwi, have rudimentary wings which are useful only for fluttering, or have quite well-developed wings on which they can glide for some distance but lack the muscles needed for flapping flight. They live in temperate or warmer climes, a necessary requirement as food is then available year-round without the need to migrate. Most of the aquatic flightless species still use their modified wings for swimming. Colder environments with seasonal temperature swings, including harsh winters and ice-bound land, were not a problem for the aquatic marine birds such as the penguins and the now extinct northern ocean species—the great auk and spectacled cormorant—as they had access to open water and constant food supplies. These birds are adapted to live in cold water, and in fact it is a necessary requirement for their survival. Most penguins range widely across the cold southern oceans all winter, often dispersing several thousand miles from their summer nesting rookeries, and not coming ashore for weeks at a time. Even though their distribution extends north into the tropics and even to the equator in the Pacific Ocean, they are dependent upon cold water to control their body temperatures and provide the best environment for their food sources. This cold water is provided by the currents that swirl along the western coasts of Africa and South America, bringing water and nutrients up from Antarctica.

Flightless birds are those that cannot sustain flapping flight, but it is more difficult to categorize birds that are quite obviously poor fliers. Many species, including several rails and the tinamous and scrubbirds, are flutterers or "hop-fliers." They run, become airborne for a short distance just skimming the ground, then touch down and repeat the process. There are many other tree-living species that barely fly, such as the pheasant-like hoatzin (*Opisthocomus hoazin*) and the Henderson Island fruit dove (*Ptilinopus insularis*), which flap heavily and clumsily between neighboring trees, and New Zealand's tiny, light-weight rifleman (*Acanthisitta chloris*), which can fly short distances but generally runs up tree trunks like a tree-creeper and then "floats" down to the base of a nearby tree. None of these birds are capable of sustained flight, but they have obviously not completely lost the power of flight, so it would be wrong to call them flightless. In many cases their degree of flightlessness is uncertain. Indeed, there are some—such as the unique birds in Madagascar called mesites and several of the secretive South America tapaculos—which have not been studied in the wild, but did not fly on the rare occasions when they were observed. Therefore, until more is known about these birds, both behaviorally and anatomically—whether in fact they have the muscles and wings capable of flight—it seems more appropriate to consider them semi-flightless.

Loss of flight is not just a recent phenomenon that has occurred in birds that took many years to develop their flying ability. On the contrary, there was a greater variety of flightless birds in prehistoric times, and among the living species both the penguins and the ratites have been flightless for millions of years, which is a sure indication of the success of their way of life. Throughout the evolution of birds and flight, many species lost their ability to fly and then succumbed to the pressures of

Galapagos Penguin *Loss of flight in the waterbirds required either the use of the legs for propulsion, or in the case of the penguins, the development of their wings as paddle-like flippers, allowing them to "fly" through the water, and requiring a keeled breastbone and powerful breast muscles.*
Photo: Courtesy Dr. Robert H. Rothman, Department of Biological Sciences, Rochester Institute of Technology

life, especially the evolving land carnivores from which they could not escape by flight. The fossil record proves that many flightless species became extinct since birds began evolving from the dinosaurs, and many of those early birds were carnivorous, and very large. They included the *Diatrymas*, which lived in Eurasia and North America during the Paleocene and Eocene Epochs, between 65 and 38 million years ago. They were flightless meat-eating birds that rivalled the ostrich in size, but had a head as large as a modern pony's and a powerful flesh-tearing bill. Later, during the Oligocene Epoch and surviving until near the end of the Pliocene 4 or 5 million years ago, cursorial or running birds called *Phorusrhacids*, which were 6 feet (1.9 m) tall, chased small mammals on the Argentine pampas. But there were also some giant vegetarian species, including those that lived in Australia from about 15 million years ago until probably just a few millennia back. These have been named the *Dromornithids*, and they ranged in size from one similar to the present-day Australian emu (*Dromaius novaehollandiae*) to the largest bird ever known, which stood 10 feet (3 m) tall and likely weighed over 1,000 pounds (454 kg). They still survived when the first aborigines arrived in Australia about 40,000 years ago, as they are included in tribal legends, and it is possible they were exterminated by them.

The world's largest contemporary birds, the giant moas of New Zealand and Madagascar's elephant birds, would still be alive today had it not been for man's direct assault on them during the past 1,500 years. Although it seems likely that flightless birds were more vulnerable to evolutionary pressure than flying birds, their increased vulnerability during historic times is beyond doubt. The many extinctions of flightless species during the last four centuries is out of all proportion to the losses of flying birds during the same period, but for once the normal evolutionary pressures were not responsible and Homo sapiens is entirely to blame.

Prehistorically, flightless birds were most heavily concentrated in the Southern Hemisphere, but until recently fifteen forms (species or subspecies) also occurred north of the equator, but ten of these vanished during the last two centuries. The great auk (*Pinguinus impennis*), spectacled cormorant (*Phalacrocorax perspicillatus*), Hawaiian rail (*Porzana sandwichensis*), and Kittlitz's rail (*Porzana monasa*) did not survive the nineteenth century. The rail of Iwo Jima Island (*Poliolymnas cinereus breviceps*) was last seen in 1924; the Syrian ostrich (*Struthio camelus syriacus*), Wake Island rail (*Gallirallus wakensis*), and Laysan rail (*Porzana palmeri*) disappeared during the mid-1940s, and the Colombian grebe (*Podiceps andinus*) and Atitlan grebe (*Podylimnas gigas*) were exterminated in the last years of the twentieth century. Consequently, the North African ostrich (*Struthio c. camelus*), the Okinawa rail (*Rallus okinawa*), Guam rail (*Rallus owstoni*), and the Zapata rail (*Cyanolimnas cerverae*) are the Northern Hemisphere's only surviving flightless birds, and those in the Southern Hemisphere fared only slightly better. Without exception humans were responsible for these extinctions through hunting, settlement, environmental changes, and the introduction of alien animals.

Several of the world's rarest birds are flightless. New Zealand's large parrot, the kakapo, and the Lake Junin grebe (*Podiceps taczanowski*) of Peru may be the rarest of all, with known populations of less than 100 individuals each. They are closely followed by four other New Zealand birds, the Campbell Island teal (*Anas nesiotis*), the giant rail known as the takahe (*Notornis mantelli hochsteteri*), and two kiwis—the rowi (*Apteryx rowi*) and the tokoeka (*Apteryx australis*)—with less than 300 specimens each.

Several rails, like the New Caledonia rail (*Galirallus lafresnayanus*), Woodford's rail (*Nesoclopeus woodfordi*), and Cuba's Zapata rail, are now rarely seen and are believed to be on the verge of extinction; but the rails are secretive birds and both the Samoan moorhen (*Gallirallus pacificus*) and the invisible rail (*Habroptilus wallaci*) were believed extinct for many years until recently, when possible sightings raised hope of their continued existence. There are perhaps 1,500 Galapagos penguins (*Spheniscus mendiculus*) still alive; New Zealand's yellow-eyed penguin (*Megadyptes antipodes*) has a population of about 2,000 breeding pairs, and the Fiordland crested penguin (*Eudyptes pachyrynchus*) about 3,000 pairs. Conservation programs have already successfully assisted some of these birds. The kakapo

Table I.1
The Living Flightless Birds

Family	Species & Endemic Insular Races	Distribution
Struthionidae	Ostrich (*Struthio camelus*)	Africa
Rheidae	Greater Rhea (*Rhea americana*)	South America
	Lesser Rhea (*Pterocnemia pennata*)	South America
Dromaiidae	Emu (*Dromaius novaehollandiae*)	Australia
Casuariidae	Double-wattled Cassowary (*Casuarius casuarius*)	Australia, New Guinea
	Single-wattled Cassowary (*Casuarius unappendiculatus*)	New Guinea
	Dwarf Cassowary (*Casuarius bennetii*)	New Guinea
Apterygidae	North Island Brown Kiwi (*Apteryx mantelli*)	North Island
	Rowi (*Apteryx rowii*)	South Island
	Tokoeka or South Island Brown Kiwi (*Apteryx australis*)	South Island, Stewart Island
	Great Spotted Kiwi (*Apteryx haasti*)	South Island
	Little Spotted Kiwi (*Apteryx oweni*)	New Zealand's offshore islands (a)
Podicipedidae	Junin Grebe (*Podiceps taczanowski*)	Peru
	Short-winged Grebe (*Rollandia microptera*)	Peru, Bolivia
Spheniscidae	Emperor Penguin (*Aptenodytes forsteri*)	Antarctica
	Adelie Penguin (*Pygoscelis adeliae*)	Antarctica
	Chinstrap Penguin (*Pygoscelis antarctica*)	Sub-Antarctic Islands
	King Penguin (*Aptenodytes patagonica*)	Sub-Antarctic Islands
	Gentoo Penguin (*Pygoscelis papua*)	Sub-Antarctic Islands
	Macaroni Penguin (*Eudyptes chrysolophus*)	Sub-Antarctic Islands
	Royal Penguin (*Eudyptes schlegeli*)	Sub-Antarctic Islands
	Rockhopper Penguin (*Eudyptes chrysocome*)	Temperate Southern Ocean Islands
	Snares Island Penguin (*Eudyptes robustus*)	Snares Islands
	Erect-crested Penguin (*Eudyptes sclateri*)	Sub-Antarctic Islands
	Fiordland Penguin (*Eudyptes pachyrynchus*)	South Island, Stewart Island
	Yellow-eyed Penguin (*Megadyptes antipodes*)	South Island, Sub-Antarctic Islands
	White-flippered Penguin (*Eudyptula albosignata*)	South Island
	Magellanic Penguin (*Spheniscus magellanicus*)	South America
	Humboldt's Penguin (*Spheniscus humboldti*)	South America
	Jackass Penguin (*Spheniscus demersus*)	South Africa
	Blue Penguin (*Eudyptula minor*)	Australia, New Zealand
	Galapagos Penguin (*Spheniscus mendiculus*)	Galapagos Islands
Phalacrocoracidae	Galapagos Flightless Cormorant (*Nannopterum harrisi*)	Galapagos Islands
Anatidae	Auckland Island Teal (*Anas a. aucklandica*)	Auckland Islands
	Campbell Island Teal (*Anas nesiotis*)	Campbell Islands
	Magellanic Flightless Steamer Duck (*Tachyeres pteneres*)	South America

(*continued*)

Table I.1
(*continued*)

Family	Species & Endemic Insular Races	Distribution
	Falkland Flightless Steamer Duck (*Tachyeres brachypterus*)	Falkland Islands
Rallidae	North Island Weka Rail (*Gallirallus australis grayi*)	North Island
	Buff Weka Rail (*Gallirallus australis hectori*)	Chatham Island (b)
	Black Weka Rail (*Gallirallus a. australis*)	South Island
	Stewart Island Weka Rail (*Gallirallus australis scotti*)	Stewart Island
	Okinawa Rail (*Rallus okinawa*)	Okinawa
	Henderson Island Rail (*Porzana atra*)	Henderson Island
	Aldabra Rail (*Canirallus cuvieri aldabranus*)	Aldabra Islands
	New Guinea Flightless Rail (*Megacrex inepta*)	New Guinea
	New Britain Rail (*Gallirallus insignis*)	New Britain
	Inaccessible Island Rail (*Atlantisia rogersi*)	Inaccessible Island
	Woodford's Rail (*Nesoclopeus woodfordi*)	Bougainville, Solomon Islands
	Samoan Moorhen (*Gallinula pacificus*)	Savaii Island
	Snoring Rail (*Aramidopsis plateni*)	Sulawesi
	San Cristobal Moorhen (*Gallinula silvestris*)	San Cristobal Island
	Invisible Rail (*Habroptila wallaci*)	Halmahera
	Auckland Island Rail (*Rallus pectoralis muelleri*)	Auckland Islands
	Gough Island Moorhen (*Gallinula nesiotis comeri*)	Gough Island
	Guam Rail (*Rallus owstoni*)	Guam
	Lord Howe Island Rail (*Tricholimnas sylvestris*)	Lord Howe Island
	New Caledonia Wood Rail (*Gallirallus lafresnayus*)	New Caledonia
	Zapata Rail (*Cyanolimnus cerverai*)	Cuba
	Tasmanian Native Hen (*Tribonyx mortieri*)	Tasmania
	South Island Takahe (*Notornis mantelli hochsteteri*)	South Island
	Giant Coot (*Fulica gigantea*)	South America
	Calayan Rail (*Gallirallus calayanensis*)	Calayan Island
Rhynochetidae	Kagu (*Rhynochetus jubatus*)	New Caledonia
Psittacidae	Kakapo (*Strigops habroptilus*)	New Zealand's offshore islands (c)

(a) The little spotted kiwi has long been extinct on North Island and is now believed extinct on South Island also, but has been successfully introduced onto Kapiti and several other offshore islands.
(b) The buff weka is now extinct in its natural range on South Island, but was introduced to Chatham Island, where it thrives. However, it was recently determined to be a color phase of the black weka.
(c) The last kakapos in their natural range on Stewart Island were moved recently to predator-free offshore islands—Codfish (Whenua Hou), Mana, Maude, and Little Barrier Islands.

and little spotted kiwi (*Apteryx oweni*) are now extinct in their natural range, and survive only on predator-free islands to which they were translocated. The Guam rail was exterminated in the wild, but both it and the equally rare Lord Howe Island rail (*Tricholimnas sylvestris*) are now thriving in captivity, and many have been returned to their ancestral homes.

The Convention on International Trade in Endangered Species of Wild Fauna and Flora (CITES), which many of the world's nations have ratified, lists several flightless birds in its Appendix I (endangered birds in which trade is prohibited) and in Appendix II (threatened birds in which trade is monitored). Surprisingly, neither the Junin grebe nor any of the kiwis are included. In the kiwis' case this is obviously because they are restricted to New Zealand where they are strictly protected, yet so are the kakapo and Campbell Island flightless teal, which are both listed in Appendix I. Similarly, the Lord Howe Island rail is included in Appendix I, yet other rare species such as the Guam rail, Okinawa rail, and Aldabra rail (*Canirallus cuvieri aldabranus*) are excluded. The only penguins covered by CITES regulations are the still fairly plentiful African or jackass penguin (in Appendix II), and the Peruvian or Humboldt penguin (in Appendix I).

Some flightless birds will never be common as their populations cannot increase beyond the territorial capacity of their small island habitats. The Gough Island moorhen (*Gallinula nesiotis comeri*) and Henderson Island rail (*Porzana atra*), for example, each number about 5,000 individuals, which are by no means large populations, but they are stable and secure on their very isolated mid-oceanic islands, unless man-made environmental changes threaten them. Similarly, the Auckland Island flightless teal (*Anas a. aucklandica*), a race of the type species living in New Zealand, numbers no more than 1,500 birds, but is believed secure due to the remoteness and inaccessibility of its sub-Antarctic island home. Therein lies the anomaly of the status of flightless birds. After losing so many insular forms to the ravages of settlement and introduced animals, islands now offer the best chance of survival for some flightless species.

1 Losing the Advantage

The origin of birds remained a mystery until the middle of the nineteenth century when the fossil of a bird-like creature was discovered in a dry lake bed in Bavaria. It was estimated to be 150 million years old and was named *Archaeopteryx*, but despite its large, feathered wings the living version of this creature was a glider, not a flier. The bone structure needed to support flight muscles was not developed, and its tiny cerebellum or hind-brain, which monitors flight, and its long and bony, stabilizing tail show that it was not equipped for the aerial maneuvering of the modern bird. Although a few prominent zoologists thought *Archaeopteryx* was the link between the dinosaurs and birds, it was generally believed that it proved ancient reptiles were the bird's ancestors, and there was no evidence to seriously challenge this belief until almost the end of the twentieth century, when bird-like fossils were discovered in a number of countries, especially China. They had several features in common with modern birds, including a wishbone, a sternum, and very primitive feathers, and showed conclusively the line of progression from the dinosaurs to birds[1] over a period of at least 120 million years. The change from ground-dwelling, tough-skinned animals into feathered flying creatures was a remarkable transformation indeed.

Flying seems such a boon, it is inconceivable that some birds would choose not to fly, but that is exactly what happened during the course of evolution, and those that did not use their wings eventually lost the power of flight. How long this reverse process took remains as much a mystery as the development of flight, but recent observations on the atrophy of the flight muscles through a bird's inability or unwillingness to fly indicate that flight can be lost very quickly—possibly in just a few generations in some species. However, loss of flight is not necessarily synonymous with loss of wings, and it is likely that the reduction of normal wings to the tiny vestigial appendages which are the kiwi's and cassowary's wings, has taken millions of years.

The development of a soft, insulating layer of feathers allowed the early birds to maintain higher body temperatures and so become warm-blooded. Feathers eventually developed into a necessary requirement for flight, although not the only one, and allowed colonization of colder environments and dispersal to otherwise inaccessible locations. Like *Archaeopteryx*, the first birds could only glide, but the eventual development of the keel and powerful breast muscles then made flapping-flight possible. Surprisingly however, insulation and not flight is the feathers' most important role. Flight is a secondary function, for although flying birds may manage quite well when they are grounded, they cannot survive a lengthy drop in body temperature unless they have evolved to become torpid daily like the hummingbirds and swifts, or to actually hibernate like the poorwill. Insulation and maintaining body temperature are the feathers' only functions that apply to all birds, and with their protection they were able to colonize the world's coldest regions and survive in temperatures many degrees below freezing for months at a time. As birds evolved, the demands of flight affected almost every aspect of their structure and physiology. They became streamlined and their muscles and bones were modified to provide maximum power and minimum weight. Their lungs and circulatory systems underwent great changes to provide the oxygen needed for such strenuous activity, and they developed a digestive system able to rapidly process the high-energy foods needed to fuel their flight. Flight also required highly developed sight and balance, and the ability to quickly control muscular activity to prevent crash landing, to alter direction rapidly, to avoid obstacles, or to chase and seize their prey.

Mallard   *A flying bird with large wings of stiff primary and secondary feathers, attached to a keeled sternum and powered by breast muscles that may account for 15 percent of its body weight, a mallard drake "explodes" almost vertically from the water when startled.*
Photo: Courtesy U.S. Fish & Wildlife Service

Birds basically have two muscular-skeletal systems fused into one rigid frame. Their hindlimbs and associated muscles, which are for walking, running, climbing, assisting take-off, and for absorbing the shock of landing, are attached to the pelvic girdle. Their forelimbs, which have developed into wings powered by large muscles, are attached to the rigid sternum, and allow rapid response to threats. In addition, their bones are mostly hollow and thin-walled, with internal reinforcing struts giving them a honeycombed appearance and making them lighter and stronger. Airspaces linked to the lungs provide a ventilation system for the muscles. The most distinctive feature of the flying bird's skeleton is the presence of a breastbone shaped like a boat's keel, which provides a greater surface for the attachment of large breast muscles, without which flight is impossible.

Wings are a bird's forelimbs modified for flight. They are concave, which is more efficient for flying than a flat surface, and comprise the same components as human hands and arms, with an upperarm, elbow joint, forearm, wrist, hand, and fingers; but unlike our arms, they rotate only at the shoulder. The major feathers used in flight—the primaries and secondaries—are attached respectively to the fingers and the forearm, and are anchored to the bones by connective tissue. The primaries are mainly responsible for control and propulsion, while the secondaries provide lift. However, as the removal of the primaries on one wing has shown—through the wing-clipping or pinioning of captive birds to make the wings asymmetrical and normal flight impossible—the primaries of both wings are necessary to power lift-off unless the bird is assisted by a strong wind.

Air is a thin medium and a bird in flight must support its weight and overcome the drag caused by its forward movement, which is achieved by pushing air downward with each wing-beat. More power is naturally required for the wing's downstroke against gravity than the return upstroke, and this is provided by large pectoral muscles situated on the breastbone and keel, and attached by stout tendons to the bird's humeri or upper wingbones, which depress the wings for the down-stroke. Other muscles, called the supracoracoideus, also attached to the keel beneath the pectoral muscles, thread through the wing's socket bones and are attached to the upper surface of each humerus. These elevate the wing on the upbeat in slow flight, but in fast flight there is usually sufficient lift to bring the wings back up without much muscle power. Small muscles also control the movement of the feathers. The muscles involved in flight are so large they comprise about 15 percent of the total body weight of a flying bird. The pressure from the wing is transmitted through the shoulder joint only, unlike the bats, whose wings are also supported by their hind legs and their flanks.

Flight is physically demanding and requires more energy than any other form of bird movement. Even during flight without wing-beats, such as soaring and gliding, muscle fatigue still occurs as air pressure tends to lift the wings, which must remain horizontal. Flight becomes more demanding as a bird's body size increases, for the wings must carry relatively more weight in large birds than small ones because the ratio of wing surface to body volume is less, which is called wing-loading. The power needed to fly is more diffcult to provide as size increases, and if a bird doubled its size and still flew in the same manner, the surface area of its wings would have to increase by at least a factor of two to support its body in the air.

The bustards, wild turkeys, pelicans, condors, swans, and albatrosses are the largest living flying birds, with the maximum flying weight in modern birds being 46 pounds (21 kg) of body weight, which is reached in the great bustard (*Otis tarda*). Such heavy birds must face into the wind and run over ground or water to reach the minimum air speed needed for lift-off, or launch themselves from a high point. When airborne they make good use of thermals, soaring and gliding to conserve energy, and reducing drag by spreading their primary feathers to form slots.

Some prehistoric species were much heavier, like the condor, which lived in South America in Pleistocene times and is believed to have weighed 85 pounds (38 kg). It would have required massive wings to support such a large body weight, and could probably only become airborne by launching off Andean peaks. Of all the living flightless birds only the large ratites—the ostrich, emu, cassowary, and rheas—plus the emperor penguin and possibly the king penguin, have grown beyond the maximum flying weight. However, large wings do not guarantee flight; muscles are also necessary. As birds like the kakapo and kagu prove, relatively large wings are useless for flight when they lack the muscles to power them. Similarly, a keel and heavy breast muscles do not indicate good flying ability, unless associated with operational wings. Penguins have a definite keel and well-developed pectoral muscles but their wings have become modified for swimming. The flightless steamer ducks also have large pectoral muscles and powerful wings—although they are small and not fully extendable—which they beat furiously to assist their passage over the water, and possibly also when underwater. Even a keel, good breast muscles, and feathered wings may not permit flight because of the structure of the muscles. The thick, white meat on a turkey's breast is mainly the product of do-mestication and selective breeding, and contains white muscle fibers that have a limited blood supply and are fueled by glycogen. Consequently, the turkey's breast is inefficient at refuelling quickly or in removing waste products, so it is useless for flight, although before it reaches full adult weight the immature domestic turkey can fly short distances. In contrast to the white breast muscle, and in keeping with its predominantly ground-dwelling disposition, the turkey has dark thighs because of the predominance of red fibers. These are richer in blood vessels and energy-storing myoglobin and burn fat instead of glycogen, allowing the muscle to contract for longer periods. Birds that are frequently airborne similarly have dark-red breast muscles to power their sustained flight.

A bird's wing muscles expand and contract the rib cage during flight so that its breathing matches the wing-beats. Some of the inhaled air goes into the lungs, but oxygen-rich air also goes into air sacs in the body. This air passes through the lungs when it is exhaled, so the bird also benefits from its oxygen content when breathing in and out. A bird's heart, which is proportionately up to four times larger than the human heart, beats very fast—up to 1,250 times per minute in the hummingbird—to carry oxygen rapidly throughout the body. The heart's four-chambered design separates the oxygenated blood pumped around the body from the oxygen-depleted blood being returned to the heart, ensuring that the cells receive a rich supply of oxygen.

Flying provided birds with many benefits. Within their home range it improved their chances of finding food and escaping predators, and it provided greater nesting security, high in trees, on sheer cliff ledges, and on remote, predator-free islands. Flight enabled birds to expand their ranges and territories, to locate food sources from a great height, and make local migrations in search of food and water. It also allowed them to make longer journeys to colonize distant habitats, to cross mountain and water barriers, and twice annually to migrate across the hemispheres to escape harsh winters and the lack of food. Flying provided many advantages over the restrictive, terrestrial lifestyle of their ancestors, with birds flying for a specific purpose, generally connected to survival, and only occasionally including what could anthropomorphically be called "flying for the fun of it." When the need no longer existed, birds flew less and less until eventually they could no longer fly. All living flightless birds show evidence that their ancestors could fly. They lost the ability simply because they did not use their wings often enough to maintain the muscles needed for flight. Or, in the case of the penguins and the extinct great auk, wings evolved into flippers for more efficient swimming.

Several birds were predisposed to flightlessness because of their habits. It is a family trait of the neotropical tapaculos, Madagascar's mesites, the Australian scrub-birds, New Zealand's wrens and wattlebirds, and the tinamous—the ancestors of the ratites. They all still fly, although weakly, and are considered semiflightless birds. With the exception of the wrens and the wattlebirds (the kokako and the saddle-backs) they reached this condition in the face of avian, reptilian, and mammalian predators, so it is ironic that the three races of the bush wren plus the flightless Stephen Island wren (*Acanthisitta lyalli*) were exterminated in recent years by alien predators.

The environment and their behavior predisposed birds for flightlessness. A bird's ability to acquire food without flying was a strong incentive for not using its wings. Birds of prey seem unlikely candidates for flightlessness, yet the long-extinct Cuban giant owl lost its wing-power through lack of use. Nesting and roosting on or near the ground also reduced the need to fly, and birds dependent upon the tree tops for their food and nest sites were obviously poor candidates for flightlessness. Potential flightless land birds also had to live in an environment that provided food throughout the year, with no requirement for long migratory journeys, which implied residence in a temperate or warmer climate, whereas the aquatic species merely had to enter the water to find their food. Birds that skulked in the undergrowth and ran back to its security when threatened were certainly predisposed to losing their powers of flight. Finally, combining this behavior with nocturnal or crepuscular activity (at dusk and dawn) gave birds an even better start on the road to flightlessness. There was, however, a big deterrent to becoming flightless—the presence of predators.

The level of predation was the most important environmental criterion for the loss of flight. Most of the living flightless birds developed in an environment free of predators, either initially in the case of the evolving large ratites, or permanently for certain birds on oceanic islands and remote lakes. When species predisposed to flightlessness were cast away on an island paradise, free of predators, many soon became flightless. The ratites of South America, Africa, and Australasia are presumed to have lost their flight before the advent of the mammalian carnivores, and with their increased size and speed were later able to hold their own. Few birds

North Island Kokako *A wattlebird or wattled crow—a semiflightless bird—the kokako is believed to be one of New Zealand's original species, "on board" when the islands drifted away from Gondwanaland long ago. With its short and rounded wings its flight is weak and labored for short distances only, yet it has not totally lost the power of flight, even after such a long time in a predator-free paradise.*
Photo: Courtesy Department of Conservation, New Zealand. Crown Copyright. Photographer: Dick Veitch

totally lost their flight while living alongside predators that were a threat to them. The Tasmanian native hen (*Tribonyx mortieri*) is one, now flightless despite the presence of marsupial carnivores, while its ancestor in Australia still flies.

Island birds have shown they are particularly susceptible to becoming flightless, and that their flying powers can deteriorate rapidly when there is no need to fly. In relation to the millions of years needed for flight to develop, its loss can sometimes be measured in decades. As a group, the rails were certainly behaviorally predisposed to losing their flight, and their unwillingness to fly was accentuated when they became established on islands. It is significant that totally flightless rails live only on islands, although two of these—Tasmania and New Guinea—are certainly not free of native predators. However, predisposition is not necessarily a requirement for flightlessness, and island life encouraged loss of flight in some unlikely birds when conditions were favorable. Although not yet totally flightless, even the tree-top-feeding Henderson Island fruit dove (*Ptilinopus insularis*) and the tiny arboreal rifleman (*Acanthisitta chloris*) of New Zealand's forests show that loss of flight may happen to any bird when predation is absent or at least at a very low level.

The pectoral or flight muscles show the first effects of the lack of regular wing exercise in land birds. They deteriorate rather quickly through lack of use, and although some flightless species, such as the kakapo and kagu, still have relatively large wings, they are inadequate for flight without the necessary musculature. Infrequent flight and the degeneration of the flight muscles eventually results in a reduction of the keel, changes in the size and structure of the wings, and reduced clavicles or collarbones. When the keel is totally reduced the chest is flat or raft-like. The wings may then be virtually absent as in the kiwis and the cassowaries, or comparatively large like those of the ostrich and rhea, which are useless for flight but are used in displays when the birds dash about in all directions flapping their wings. The keel is so important for flight that in fossils, where the extent of a bird's soft tissue such as the pectoral muscles can no longer be determined, the shape of the sternum indicates whether flight was ever possible. In most aquatic species the loss of flight had different results because the wings were used for swimming. Penguins' wings were modified as flippers, which required the same or perhaps even more musculature than flying birds for moving in the water's denser medium, so they have a well-developed keel and pectoral muscles. Penguins are therefore the most specialized of all flightless birds, for practically all others merely lost the use of their wings and in many cases grew larger. The flightless steamer ducks are an anomaly, for although they cannot fully extend their small wings, powerful muscles enable them to complement their large feet as they "steam" through the water and swim beneath the surface.

The loss of flight removed all constraints on body size, whether birds were terrestrial or aquatic, and so many species grew larger that flightlessness is now associated with increased size, although there are still many small flightless birds. In fact, flightless species have the greatest size range of all birds, from the ostrich, which stands 8 feet, 3 inches (2.5 m) high to the tiny Inaccessible Island rail (*Atlantisia rogersi*), which is only 5 inches (12.5 cm) long. Airborne flight has limited the size of the living auks—small, stubby birds with short wings—but when the ancestors of the great auk (*Pinguinus impennis*) became flightless and this size constraint was removed, it had reached a weight of 44 pounds (20 kg) when it was exterminated in the nineteenth century. The emperor penguin (*Aptenodytes forsteri*), the largest member of the family, now weighs 66 pounds (30 kg) and prehistorically there were even larger species of penguins. The flightless steamer ducks are larger and bulkier than their flying ancestor, which still lives alongside them. The size increases in land birds were even more dramatic, reaching their peak in the recently extinct moas and elephant birds, and in the living ostriches, cassowaries, emu, and rheas. Several other New Zealand birds actually grew larger than the kiwis when they stopped flying. They included the turkey-sized takahe (*Notornis mantelli hochsteteri*), which is the largest living rail, and the kakapo (*Strigops habroptilus*)—the largest living parrot—and a recently extinct goose which stood almost 39 inches (1 m) tall.

The other major morphological changes in flightless birds were increases in the size of their feet, legs, and leg muscles, in keeping with their terrestrial way of life and to support a heavier body. In living birds this has reached its peak in the large ratites with their thick legs and tremendous thigh muscles, and the ability to inflict serious and even fatal injury through kicking. It has also provided the means to outrun their potential predators, at speeds which no other land bird, and few mammals, can match.

Little Spotted Kiwi *A flightless bird, the kiwis are very ancient birds whose ancestors were probably on New Zealand when it drifted away from Gondwanaland millions of years ago. They are the only survivors of the greatest concentration of ratites or cursorial birds ever known, most of which succumbed to the pressures of human colonization. Kiwis show the greatest wing reduction of all flightless birds, their vestigial wings being only about 2 inches (5 cm) long, and hidden by their feathers.*
Photo: Courtesy Otorohanga Kiwi House

An increase in body size is generally an indication of how long a bird has been flightless, and most small rails on isolated islands are believed to be fairly recent colonists, although there are several exceptions. One is the Inaccessible Island rail (the world's smallest flightless bird), which has lived on its small mid-Atlantic island for so long that its ancestry and origins are uncertain. Then there are the kiwis, which have not increased in size like the other ratites despite their many years in New Zealand. This has been attributed to their nocturnal and secretive lifestyle, possibly to avoid the predatory native diurnal birds that are now extinct, and perhaps because they occcupied a niche that did not conflict with the larger moas, a very successful group of flightless birds, until the Polynesians arrived.

Flightlessness and predation are undeniably connected, and birds living on islands or remote lakes, where predation was nonexistent or very low, became flightless because the major reason for using their wings was absent. However, there are a number of puzzles. Why, if predation was the control, have several birds on the continental land masses reached the semiflightless stage in the face of numerous predators? Also, why did birds become flightless on islands with predators which could certainly be expected to prey upon them. Obviously, it is not solely the lack of predators that reduces a bird's need to fly, as is so often stated. While no predation is optimum, for some birds even the presence of predators did not force them to use their wings sufficiently to maintain them, and they have lost or almost lost their use. The only possible conclusion is that it was the level of predation, not the lack of predators, that produced flightlessness.

Note

1. This line of progression did not include the Pterodactyls or flying reptiles, which became extinct about 75 million years ago. They were unrelated to the developing birds and flew by means of a membrane connecting the forelimbs and hindlimbs, similar to modern-day bats.

2   The Continental Drifters

The large flightless birds called ratites are the giants of the bird kingdom and some that lived in Madagascar and New Zealand until quite recently were the largest species ever known. The survivors, which include the seven heaviest birds on earth, all live in the Southern Hemisphere, with just one extending its range north of the equator in Africa, and until recently also into the Middle East. They are cursorial or running birds, which inhabit all the world's great grasslands and drier regions—the African veldt, savannah, and desert; the Australian outback; and South America's pampas,[1] campos,[2] and altiplano.[3] Several species live in rain forests, tropical ones in northern Australia and New Guinea, and the temperate forests of New Zealand. The large species are diurnal, while the smaller ones are nocturnal or active at dusk and dawn.

Before the theory of continental drift was understood, the ratites were believed to have developed their similar characteristics, even on such widely dispersed land masses, as a result of convergence—when unrelated birds develop similarities and adaptations as a result of having the same lifestyle. It was also thought that their ancient ancestors were flightless, and that their primitive characteristics, such as rudimentary wings and vestigial flight feathers, indicated an early phase of bird development directly from their dinosaur ancestors. Ratites originated about 100 million years ago, so are certainly very ancient birds, but recent comparative studies of their structure and behavior, and that of the secretive, dull-plumaged, chicken-like South American birds called tinamous, suggest they share a common ancestor. The ratites have the same very primitive palate bone structure, unlike any other living birds, which indicates probable descent from the tinamous. In addition, ratites also have several features that are characteristic of flying birds. They have a pygostyle, which is the fused end of the vertebral column that supports the fleshy tail; the ostrich and rhea have an alula or bastard wing, which is the first digit of the wing to which are attached several flight-covert feathers; and fused "hand" and "wrist" bones, all of which are associated with flight. All of this is considered more than sufficient proof that they evolved from flying ancestors and acquired their unique way of life rather early in the evolution of the birds.

Until recently, the ratites were said to belong to the group or superorder *Acarinatae* (no *carina*, Latin for keel), which is one of the three major groups in the bird kingdom, the Class *Aves*. The others are the penguins, which are in a group of their own, and then the many flying birds that have a sternum shaped like a boat's keel and belong to the subclass *Carinatae*. Ratites are characterized by their large

Ostrich   *A male ostrich, the largest living bird and therefore the largest flightless bird, is also the fastest bird on land. Now restricted as a wild species to sub-Saharan Africa, but occurring ferally in central Australia, and farmed commercially in many countries for its eggs, meat, skin, and feathers. The ostrich reaches a height of 9 feet (2.7 m) to the top of its head and may weigh 340 pounds (154 kg).*
Photo: Doxa, Shutterstock.com

bodies, powerful legs, and the great reduction of their wings (in some species) and flight muscles. Consequently, the keeled sternum to which the breast muscles were formerly attached was no longer needed, and they became flat-chested and are said to have a raft-like sternum (hence ratite, from *ratis*, Latin for raft). However, the name *Acarinatae* is now considered by some taxonomists to be inaccurate for the ratites, as their supposed ancestors the tinamous have keeled sternums and can still fly, although not too well. So it has been suggested that a more appropriate name for all birds excluding the penguins would be *Paleognatae* for those with primitive palates whether they fly or not, thus joining together the tinamous and ratites; and *Neognatae* for the flying birds with more modern palate bones. This last group would include birds that have lost their flight more recently than the ratites, mainly as a result of insular isolation.

The ratites are currently divided into three groups—the ostrich, the two species of rheas, and the group that contains the emu, cassowaries, and kiwis. The tinamous are considered their closest living relatives. Their evolution is not exactly clear-cut, but there is no doubt that it is entwined with the breakup of Gondwanaland. Acceptance of the continental drift theory, that all land was once one large mass from which the present-day continents drifted away, confirms South America as their center of origin and makes their evolution from a common ancestor more understandable. Continental drift also explains why they live only on the southern continents, thousands of miles apart, for it is now recognized that Earth's land areas have been considerably redistributed over the geologic epochs. The continents as we know them were once joined in one enormous land mass known as Pangaea, which separated into Gondwanaland in the south and Laurasia in the north between 250 and 200 million years ago, long before the ratites evolved. The southern supercontinent of Gondwanaland then itself began to break up in the middle of the Cretaceous Period about 100 million years ago. Separating from Africa, South America drifted west and Australia and Antarctica moved south. They remained connected for several million years, but during this period the segment of land that became New Zealand broke away from Australia and has remained isolated ever since. Australia eventually separated from South America after being colonized by the early ancestors of the present-day marsupials, but fortunately before the carnivores evolved. Australia and Antarctica remained connected until the beginning of the Cenozoic Period about 65 million years ago. Then, approximately 35 million years ago, the Americas, which had separated during the original split of Pangaea, became reattached: India moved north to join Asia, the violence of its welcoming impact thrusting up the Himalayas; and Antarctica drifted down to the South Pole. The Antarctic continent was a lot warmer then, having been glaciated for only about 10 million years.

Flightless land birds certainly cannot cross oceans, and the present distribution of the ratites supports the theory of continental drift. Gondwanaland was obviously still intact when the ratites began evolving from their tinamou-like ancestor in South America, so the early cursorial birds were able to reach Africa, Madagascar, and Australia over land, and were present on the southern continents when they broke away. It is believed that the ancestors of the kiwis and moas reached New Zealand about 90 million years ago via Australia, and a later wave of bird colonists brought the primitive ancestors of the emu and cassowary to Australia via the Antarctic. The ratites of Australia and New Zealand are the only living birds in those countries that are clearly remnants of the ancient connection between South America, Antarctica, Australia, and New Zealand. However, it has been suggested that the rheas may be more closely related to the ratites of Australasia than to the African ostriches, implying that the rhea's ancestors actually reached South America from Australia via Antarctica long ago. All the living species of ratites are therefore examples of adaptive radiation, as their predecessors evolved into different but similar forms to suit the environments developing on the newly isolated land masses.

The past 1,500 years have shown how vulnerable flightless birds are to predators and unnatural environmental change, and many ratites became extinct then. Yet prior to this they managed to survive and evolve for millions of years on the southern continents. On the open grasslands of Africa and South America they either increased in size before the advent of the mammalian carnivores, or they evolved alongside them; their size, speed, and specialized breeding habits enabled them to survive in an increasingly hostile world. The largest flightless birds of all, the elephant birds, survived until recently because of the absence of major land predators on Madagascar until humans arrived. Similarly, the cursorial birds of Australia and New Zealand were safe because both split from Gondwanaland before the age of mammalian carnivores such as the dingo, a more recent introduction by the aborigines. The emu, cassowary, and kiwi have survived there until the present day; but the moas, like the elephant birds, were unable to cope with humans.

■ THE LARGE RATITES

Form and Plumage

In all the ratites except the kiwis the loss of flight was accompanied by a great increase in size as the weight limitations on becoming airborne no longer applied. The extinct moas and elephant birds were the largest recent[4] birds, followed by the living ostrich, cassowary, emu, and rhea. The seven species that make up these four families range in size from the ostrich at 9 feet (2.7 m) tall and weighing up to 340 pounds (154 kg), to the lesser rhea which is small by comparison, being only about 4 feet, 3 inches (1.3 m) tall and weighing 50 pounds (22.7 kg), but still well beyond the size barrier of 37 pounds (16.8 kg), which is the load limit for modern flying birds.

After their size, the most noticeable features of the large ratites are their powerful, scaly legs, their reduced wings, and their feather structure. Although they are useless for flight, both the ostrich and rhea have surprisingly large wings, and well-developed muscles in the forearm and hand. The ostrich actually has sixteen primary feathers on each wing, six more than the average flying bird. Both the ostrich and rhea employ the broken-wing tactic of the flying birds when attempting to draw predators away from their nests or young, further proof that they evolved from flying birds. The cassowary and emu have rudimentary wings and vestigial feathers, which in the cassowary are long, bare quills. Of all the large ratites only the ostrich has a tail, with long plumes; the others have no tail feathers differing from their body

Ratites Feet   *The ostrich (top) is the only bird with two toes, with a nail on the largest one only. Kicking forward, as its legs do not bend backward, it has killed people with a blow from its feet. The emu (bottom), like the other large ratites—the rhea and cassowary—has three large, nailed toes on each foot, also capable of inflicting serious injury.*

Photo: Clive Roots

plumage. All the ratites lack a uropygial or oil gland, which it is believed all birds[5] originally possessed, even the rain forest–dwelling cassowaries where waterproof feathers would appear to be essential. Despite this, the grassland ratites freely enter water and bathe.

Although they are useless for flying, the large wings of both the ostrich and rheas are useful for other purposes. They use their wings to help maintain their balance when running fast, and when twisting and gyrating during their courtship displays. Their wings are involved in their very effective escape behavior, assisting their balance when they dodge quickly from side to side; and to show dominance the male ostrich holds his head up and lifts his wings and tail feathers, and when being submissive he holds his head down and droops his wings and tail. To attract a hen, he rests on the ground and shakes his wings alternately while raising and lowering his tail. The ratites have drooping body plumage, which is quite coarse and hair-like in the emu and cassowary, which have enlarged after-feathers so that each feather is actually double plumed. In contrast, ostrich and rhea feathers lack barbules and are soft and filmy. The cassowary has a featherless head and upper neck, but its lower neck and thighs are feathered. Both the rhea and emu have feathered heads, necks, and thighs, but the ostrich's thighs and flanks are bare and it has only short down feathers and fine bristles on its long neck and small head.

In contrast to the flying birds with their well-developed wing muscles and small leg muscles, the large ratites have huge leg muscles and long, powerful legs, both to support their weight and for defense and escape. The ostrich is the only bird with just two toes on each foot, the largest of which has a nail. The other ratites have three toes on each foot, also with large nails, which reach their peak in the cassowary where the one on each innermost toe is very sharp and reaches a length of 5 inches (13 cm). They are all potentially dangerous birds, especially the cassowary and ostrich which have killed people with a slashing downwards kick. The cassowaries are without doubt the most dangerous birds in the world, and are so aggressive, temperamental, and territorial that pairs in zoos must be separated most of the time. Even emus are capable of inflicting serious injury, as shown by an adult male at the Dudley Zoo that killed its mate by inflicting many long and deep slashes across its back and sides.

Increased size has also been accompanied by an increase in speed, and the ratites are the fastest birds on land. Their powerful legs carry them quickly away from danger, which aided their survival in South America and Africa alongside the evolving mammalian carnivores. The ostrich is the fastest, with strides covering 11 feet (3.35 m), a high speed of 50 mph (80 kph) for short distances, and the ability to run for 30 minutes at 30 mph (48 kph). On rough ground its powerful feet send stones flying backwards, which caused Pliny the Elder, in his encyclopedic *Natural History*, to state that it hurled missiles at its pursuers. The emu is almost as fast, reaching a maximum speed of 25 mph (40 kph), and quickly vanishes among the outback scrub. The rhea almost matches the emu for speed, and although the cassowary is impeded by its dense jungle habitat, it crashes off through the undergrowth when startled. Normally, however, a quick glimpse is the usual view of a wild cassowary as it either disappears silently or crashes away into the bush, depending upon the perceived threat.

The ratites are long-lived, reaching their maximum in the ostrich, with a lifespan of about forty years. Longevity is not necessarily dependent upon size, however, since many birds—eagles, condors, cranes, pelicans, cockatoos, and macaws, for example—may live twenty years longer than the ostrich. Size certainly makes a difference where voice is concerned, and the large ratites have large voices. The ostrich makes a sound that David Livingstone likened to a lion's roar, and the rhea's low roaring call is more like that of a mammalian carnivore than a bird. The cassowary booms and croaks, and the emu's calls range from booming and bubbling sounds to a deep drumming.

Predation

The ostrich suffers the greatest natural predation of all the ratites. Lions and leopards are known to attack adult birds, and it is likely that adults, or at least large juveniles, are also taken by spotted hyenas, African hunting dogs, and possibly cheetahs. Chicks are naturally fair game for the same predators, plus a host of smaller carnivorous animals including jackals, caracals, servals, and eagles, and the Egyptian vulture throws stones at the eggs to open them. The chances of success for ground-nesting flightless birds in such a hostile environment appear very slim, but ostriches have developed several survival tactics. They are watchful birds, aided by their height and keen eyesight, and they associate with the great herds of mammalian grazers such as zebras and wildebeeste, which give warning of the approach of predators so that both birds and mammals benefit from the association. Nesting is obviously a very vulnerable time, but the risk of detection is reduced by the dull-plumaged hen incubating during the day, unlike the other large ratites whose males incubate the eggs unassisted. The loss of eggs and chicks is also countered by the large number of eggs that are laid and incubated. Rheas also have to contend with many predators throughout their life cycle, including maned wolves, foxes, jaguars, cougars, and the smaller pampas cat and mountain cat. These threats are countered by speed, watchfulness, large clutches of eggs, and the advantage of the males undertaking all the incubation and raising duties, leaving the hens free to lay many eggs in several nests. To draw predators away from his nest or chicks, the male rhea, like the ostrich, pretends he has a broken wing and is at the predator's mercy.

The emu has fewer natural predators but is still a very wary bird. The wedge-tailed eagle (*Aquila audax*) is probably the only natural threat to emus of all ages, but young chicks are vulnerable to the introduced dingo and red fox, and to the large monitor lizards called goannas. If eggs are left untended they are eaten by the black-breasted buzzard kite, a large raptor of Australia's interior and north, which "bombs" the eggs with stones to crack them. The aborigines claim that the kite drives the incubating bird off its nest by dive-bombing it before it begins bombarding the eggs. The cassowary's shape, drooping plumage, lack of wings, and bony head casque enable it to run fast through dense vegetation without becoming entangled, which is its first line of defense when threatened. Although it was long believed that the casque was developed for this purpose, it has recently been suggested that its main role is as a secondary sexual characteristic—to show dominance and intimidate rivals. Adult cassowaries in northern Australia have no natural enemies, and their major losses occur while crossing roads at night, but their chicks are vulnerable to large pythons and monitor lizards. In New Guinea an adult male cassowary is more than a match for the dingo-like New Guinea wild dog attempting to seize his chicks; and pythons and lizards, especially the huge and very aggressive Salvadori's water monitor, are the cassowaries' main enemies there, plus the large, forest-dwelling New Guinea harpy eagle.

Association with Humans

The ostrich has also had a long association with people, beginning in ancient Egypt. Queen Arsinoe II (the wife of Ptolemy II and joint ruler of Egypt in 279–270 BC) is recorded riding a saddled ostrich, and with their history of attempted animal domestication it is quite possible that the Egyptians bred and raised them. The ancient Romans also had a close association with the ostrich, which they trained to pull chariots, and their plumes graced centurions' helmets. They were slaughtered in the arena and were eaten afterwards; their oil was applied to the skin for cosmetic and medicinal purposes.

Subsistence hunting by primitive humans for their own use is virtually a form of natural predation, at least when they evolved alongside the prey and the hunting was sustainable by the wild populations, as it generally was long ago. This obviously excludes the excessive hunting practiced by the first colonists in Madagascar and New Zealand, who arrived to find very large birds that had evolved in the absence of humans. In both countries the killing was not sustainable and brought about the downfall of many species of giant flightless birds. The ostrich has traditionally been hunted by Africa's savannah and semidesert tribes for their own use, nowadays by the few remaining hunting and gathering societies like the Bushmen or San. The eggs of the southern race of ostrich are eaten by most San, but are taboo for the very young and elderly in some bands. They shake the eggs first and leave those that appear to contain well-developed embryos, but to increase their take they may observe a nest for several days to make sure all the eggs have been laid. Ostrich eggshells, filled with water and sealed with a grass plug and marked to denote ownership, are stored in their shelters and along hunting trails for use during the dry season. The San also catch chicks that the children use for archery practice before they are eaten, and they bring down adult ostriches with arrows tipped with acokanthera poison.

The cassowary is the largest wild land animal in New Guinea and is an important source of food for the native people, who also use its wing-feather shafts as nose ornaments, and adorn their pierced ears with cassowary bone spatulae. Chicks are caught and raised in enclosures for their meat and feathers and for use as a bride price, but the bird has never been domesticated. It is an important commodity that has been traditionally traded and exported to outlying islands, and the cassowary populations on Ceram and New Britain are believed to originate from imported New Guinea birds.

Australia's aboriginies similarly prized the emu for its eggs and meat, and rock paintings of the bird at least 3,000 years old can be seen in the Flinders Range. They also attached spearheads to the shafts with emu leg sinews, and with its feathers and blood they made the "shoes" that medicine men wore when relentlessly pursuing their victims to avenge a crime. But the emus' speed and agility in the scrub made them difficult prey, and running at speeds up to 25 mph (40 kph) and dodging between bushes they were virtually impossible to chase down; with no dogs to help them the hunters had to resort to more sophisticated methods. They

Emu *The second-tallest flightless bird after the ostrich, the emu stands 6 feet (1.8 m) tall to the top of its head, and averages 85 pounds (34 kg) in weight. Its primitive ancestors are thought to have colonized Australia from South America when the two countries were linked by a warmer Antarctica.*
Photo: Clive Roots

exploited the birds' inquisitiveness which unfortunately usually overcomes their natural wariness, and they enticed them into corrals by waving bunches of emu feathers from behind the shelter of rocks or bushes. Hunters also donned emu skins, holding the heads high with a stick, to approach and spear curious birds. They drugged emus too, adding the chewed leaves of the shrub *Duboisia hopwoodii*, which contains a narcotic, to their drinking holes, or mixed the pulp with wood ash and baited their feeding places, making them easier to approach and spear. However, the emu does not rely on speed alone to escape. Its long, drooping feathers resemble grass, and when adopting cryptic postures, such as standing still with its head held high in long grass, or lying with its neck outstretched on the ground when in the open, the color and form of its feathers provide effective camouflage.

Hunting by modern man has no relevance to subsistence hunting, and has only been sustainable when populations are well managed and a restricted annual cull is carefully controlled. There has never been such control in the hunting of any of the large ratites, and hunting for sport, food, and commerce, combined with settlement and farming, has exterminated several forms and considerably reduced the populations of others. The emus of Kangaroo Island, King Island, and Tasmania are all extinct. The Syrian ostrich was exterminated through indiscriminate shooting; the North African ostrich is now absent or very scarce throughout its range, and the South African ostrich survives as a wild bird only in protected areas. The lesser rhea is now very scarce in the northern parts of its range, and New Guinea's cassowaries suffer from increased hunting pressure as logging and mining roads open up their formerly virgin habitat.

All the grassland ratites except the lesser rhea have been domesticated and farmed commercially for their eggs, meat, oil, skins, and feathers. Emu skin produces a very high-quality leather and the skins of captive birds are superior to those of wild emus, which suffer damage from the environment and during mating. Emu ranching, popular in Australia for many years, has recently experienced a boom in North America, with an increasing market for meat, leather, and oil. There are believed to be 1.5 million emus currently being ranched in the United States, and in 1995 the U.S. Department of Agriculture approved meat inspection for all four grassland ratites. However, it is the ostrich, which is the most commercially valuable ratite, that has been the subject of several farming booms. Although domesticated for several hundred years, the first great farming boom did not begin until the end of the nineteenth century, to supply plumes for adorning hats in the western world's fashion centers. Early in the last century, 750,000 ostriches were farmed in Cape Province alone to meet the demand for plumes, and a good breeding pair sold for $400. Their feathers, which were usually plucked once annually and occasionally three times in two years, ranked fourth in total export value from South Africa. The ostrich industry collapsed during the First World War, but began again in the mid 1940s for meat and skin production. "Everywhere I looked I saw ostriches," wrote travel writer H. V. Morton after visiting the ostrich farming center of Oudtshorn in Cape Province in 1947, when their dried meat sold for one shilling per pound and skins for five shillings each. Feathers were stored in the hope that there would be another boom, and Morton was told that when it happened, "there are enough fine feathers stored away to clothe nearly every woman in the world from head to foot." The eagerly awaited feather boom never happened, although in recent years ostriches have become hot items in the exotic animal ranching business in North America, at least in the acquisition of breeding stock, with an adult pair worth $20,000 in the early 1990s.

Ostriches were also farmed in North Africa and South Australia. The North African race was imported into South Africa to improve the birds farmed there and the hybrid offspring were then exported to South Australia. Birds of this stock escaped in Australia and in South Africa and feral hybrid ostriches now exist in both countries. Although the ostrich breeding boom spread to many countries, it remains a major industry in South Africa, with an estimated 600 farmers, many near the original ostrich center of Oudtshorn, exporting meat to many countries. In August 2004, however, an outbreak of avian influenza on ostrich farms near Port Elizabeth resulted in a ban on imports by the European Union and other countries, leading to an estimated market loss of $16 million. These domesticated ostriches differ genetically from all the wild birds of course, as they have been selectively bred for specific commercial reasons, including size, food conversion and growth rate, egg-laying capability, oil production, and quality of their skins for leather production.

The large ratites are popular zoo exhibits, and have been kept and bred with varying degrees of success. The greater rhea is a commonly kept zoo bird that reproduces regularly, and its eggs have been hatched artificially in the incubator and naturally by the male. The lesser rhea is a far rarer species in zoological gardens, and has seldom bred. Surprisingly, for a species that ranges high into the Andean alti-plano, it is not as hardy in the zoo as the greater rhea, which can withstand quite low temperatures. Emus are easy to maintain and breed regularly in captivity, most eggs being artifically incubated. Most ostriches in zoos are of unknown subspecific origin or the result of past hybridizations. Of the specimens whose ancestry can be traced, the South African ostrich is the most frequently kept pure race, followed by the Masai ostrich and the North African ostrich, with the Somali subspecies being the least exhibited form in zoo collections. Only the southern or double-wattled cas-sowary is kept with any frequency in zoos and bird gardens, and has been bred on a number of occasions. The northern or single-wattled cassowary is seldom seen in collections and the Bennetts or dwarf cassowary is an even rarer bird.

The captive care or husbandry of the large ratites poses two major problems. First, their powerful build and aggressive, potentially dangerous natures, both to conspecifics and to staff, require special considerations for housing and care, par-ticularly for the ostrich and cassowary. Second, their health, which requires special care of their large legs. Adult ostriches have suffered from the lack of adequate exercise during winter confinement in northern zoos, and males have dislocated their legs during their violent gyrating. The young, growing ratites are most afflicted, however, with their legs bowing and twisting as a result of mineral and vitamin deficiencies or imbalances coupled with rapid growth and insufficient exercise. These problems have been corrected with multivitamin injections when not too severe, but slow growth and careful attention to diet is essential when raising young ratites.

Reproduction

The ratites differ from all other birds except the ducks, geese, swans, and tinamous, in having a small erectile penis. They also practice a rare form of polygamy called polyandry, which is a reversal of the normal role of the sexes in incubation and chick raising. It occurs when a female mates with more than one male in a breeding season, and the male incubates the eggs and raises the young, allowing the hen to find other mates and lay eggs in their nests. There are several variations to this theme. When the male has just one partner at a time it is known as successive poly-andry, which occurs in the cassowary and sometimes in the emu, although in the latter's case, after the female has left the nest duties to the male, she may not mate again. Harem polyandry is another form, when several females lay eggs in a male's nest, then go off to find other partners, leaving each male to incubate the eggs and raise the young. This happens in the greater rhea, in which the male may have a harem of up to fifteen females. A slightly different form of harem polyandry is practiced by the hen East African ostrich, which may be one of several hens to bond simultaneously with a male. However, only the senior hen assists him with the incubation and chick rearing, and after laying their eggs the others go off to find another mate. The southern ostrich has a further variation of this custom, in which the dominant hen helps with the incubation, but the others stay around to help raise the chicks. The tinamou, the supposed ancestor of them all, practices successive polyandry. A female lays a clutch of eggs in a male's nest and then searches for another mate, leaving each one to incubate the eggs and raise the chicks. Polygyny is the name given to polygamous behavior by the male bird, when he mates with as many hens as possible, leaving them to nest and raise the chicks. It is a more common condition, especially in species with precocial young such as the gallinaceous birds.

The female emu usually has just one mate, but she may be polyandrous occasionally and have primary and secondary mates. After leaving the primary male to incubate the first clutch of eggs she will then actively seek another unpaired

Greater Rheas *Rheas are the South American ratites, the only species to have evolved in the natural range of their ancestors, the tinamous, and the smallest of the "big four," after the ostrich, emu, and cassowary. The common or greater rhea occurs in the wooded grasslands and scrub of eastern Brazil and then southward through Uruguay to the Argentine pampas.*
Photo: Clive Roots

male who will then incubate her second clutch, and females have been seen to fight over an unpaired male. After laying her eggs she has no further role in the process of reproduction. Nesting between April and November, the male emu makes a flat bed of leaves and grasses, usually beneath a tree or large shrub, and placed to give a clear view of the surrounding country. Up to fifteen dark gray-green eggs are laid, initially at four-day intervals, then every other day. Observation on captive emus has shown that the male begins incubation after the first egg is laid, and the hen lays further eggs beside him, which he then pulls under his body without leaving the nest. In fact, for the whole fifty-six-day incubation period, the male rarely leaves the nest, not even to eat or drink, and can lose up to 22 pounds (10 kg). He turns the eggs up to four times each day, is very aggressive, and only leaves his eggs in the face of a serious threat. Cheeping noises can usually be heard through the eggshell after fifty days, and the chicks are brown with black-and-tan stripes when they hatch. They retain this juvenile plumage for four months, and are cared for by the male for another two months.

The greater or common rhea is sexually mature at the age of eighteen months, but may begin laying six months earlier. For most of the year it roams in flocks of up to thirty-six birds, but when breeding begins in September a male displays to several females, and chases away other males that enter his territory. He scrapes a shallow hole, usually in a dry area protected by bushes, and lines it with grass. Six hens usually lay their creamy-yellow eggs in his nest or next to it, but up to fifteen have been observed doing this; and he rolls the eggs beneath him with his bill. Over 100 eggs may be laid at a nest site, so the wastage is very high as the male can only incubate 25 eggs. While the male is busy incubating the eggs, the female rheas have found other nesting males and during the season may each lay up to forty eggs in several nests. The male rhea eats very little during the incubation period, rarely leaves the nest, and like the male emu, loses a lot of weight. The chicks hatch after an incubation period of thirty-eight days, and are gray with dark stripes. Three days after the eggs began hatching the male leaves the nest with his chicks, leaving many unhatched eggs behind. The chicks that hatched first have not eaten up to then, having survived on the remains of the egg yolk in their stomachs, and when the hungry male begins to feed ravenously, they follow his example.

The male ostrich scrapes out his nest depression and then begins his courtship display to attract females. The first hen to accept him lays up to ten large, cream-colored eggs at two-day intervals, while the male mates with other hens who also lay their eggs in his nest. When incubation begins the dominant first hen who will share the duties with the male pushes the surplus eggs out of the nest (although seldom her own, which she seems to recognize), leaving only about twenty eggs, which she and the male can safely cover. The drab-plumaged and less conspicuous hen incubates during the day, although the male stays nearby to help protect the nest.

The nesting behavior of the grassland ratites has undoubtedly assisted their survival. The males are fully involved with incubation and chick rearing, solely in the case of the emu and rhea, allowing the hens to lay more eggs in other nests. The large number of eggs incubated, twenty by the ostrich, fifteen by the emu, and twenty-five by the rhea, generally results in a large crop of chicks, despite the many unhatched eggs left behind in each nest. The ability of chicks to call in the egg, and for the less-developed embryos to then accelerate their development to permit hatching synchronization, allows the male to leave the nest with the majority of his potential brood.

The cassowary is the only forest-dwelling large ratite and consequently lives in conditions of much higher humidity than the others. Its nest on the forest floor is a flat bed of grass and leaves, and after laying her four to six light-green eggs in July or August the female leaves the work of incubation and chick raising to the male. The brown-striped chicks hatch after an incubation period of about fifty days and are cared for by the male for almost a year. They lose their stripes at the age of three months, but do not attain their adult plumage until they are five years old. Unlike the grassland ratites, developing cassowary chicks do not call in the egg to synchronize their hatching, which is obviously less important when so few eggs are involved.

Diet

The large ratites are mainly vegetarians. They eat seeds, fruit, leaves, flowers, and succulents, and consequently have the long digestive tracts characteristic of vegetation eaters. Like most fruit and seed-eating birds, they are all more insectivorous when young, but may have a greater liking for animal protein than we appreciate, if the sparrow-eating by zoo rheas and a captive cassowary's liking for starlings is typical. Birds' stomachs have two parts, the proventriculus or glandular section (which is generally called the stomach), where food is exposed to digestive secretions from the stomach lining; and the ventriculus or gizzard where the food is broken down by muscular action, helped by the stones and grit swallowed especially to aid the process. The ostrich's habit of eating stones originally caused much consternation before its purpose was realized. The size of each section depends upon a bird's diet. The ostrich, emu, and rhea have well-developed gizzards because of the more fibrous nature of their food, and the grit they eat goes directly there through the crop. It is not passed out with the digested food, so the stomach can be quite empty of food while the gizzard still retains its grit. However, too much grit can cause problems even for an ostrich. There are recorded cases where the grit provided to young zoo ostriches was inadvertently mixed with their food daily instead of being given separately for them to eat ad lib—as they required it—and they died when a large amount of grit compacted in their gizzards. Young captive grassland ratites have also died through eating their bedding, especially wood shavings, which can become impacted in their digestive tracts. The ostrich's food initially collects at the top of its esophagus as a single bolus, giving the impression that it has swallowed a tennis ball, which then moves slowly down to the stomach.

The cassowary has a very flabby stomach, the least muscular of all the ratites, because its diet is mainly fruit. Its breeding season coincides with the main fruiting of the trees within its range, and it is an important seed dispersal agent for forest trees. It is believed that some seeds may require passage through the cassowary's gut, to wear down their shell before they can germinate, and that the bacterial content of the dung which contains the seeds is essential for the trees' early growth.

Cassowaries are also known to eat soil for its mineral content, and captive birds have swallowed and passed wood shavings easily, unlike the grassland ratites. The cassowary's bony casque, absent in the other ratites, is believed to have evolved to protect the bird as it plunged through the dense undergrowth. However, recent observation of a captive bird using its casque to push earth aside in an apparent search for food has led to speculation that this may be the appendage's real purpose. The casque may also be associated with territorial dominance.

The cassowary, with ample fresh vegetation and fruit available in its rain forest habitat, is not confronted with the difficulties of avoiding overheating and finding sufficient fluids. This is a major concern for the grassland ratites, especially the ostrich and emu, but both are well-adapted to their environments. In the Australian outback the emu relies upon its food to supply moisture and seeks shade at midday. It has benefitted from the artesian wells and water troughs provided for domestic stock and drinks copiously when it can. Heat is not a problem for the ostrich either, as it lives in some of Africa's most desolate regions. It ranges into the central Kalahari Desert where there is no standing water for several months at a time, and even occurs in the Horn of Africa's Danakil Depression, which English explorer Wilfrid Thesiger called "country as dead as a lunar landscape." It also drinks heavily when it can, but for long periods must rely on its food for moisture, especially succulents like the monkey orange and spiny melon. Ostriches have a high tolerance for dehydration and their body temperature can increase several degrees without harm. They regulate their temperature by dissipating excess heat from all the bare areas of their bodies—their sparsely feathered heads and necks and their large expanses of bare thigh, which are exposed by raising the wings. Ostriches can also drink the briny water of the salt pans, as they have a large nasal gland, like the seabirds, that excretes excess salt.

The Species

Ostriches

The ostrich is the largest living bird, with males reaching a height of 9 feet (2.74 m) and weighing up to 340 pounds (154 kg); the females are slightly smaller. It is immediately recognizable, with its small head, long neck, large body, and sturdy legs. Its feathers lack barbules and are therefore very soft. Adult male ostriches are black and have white primary and tail feathers, and the females are gray-brown or gray. The four living races differ mainly in the color of their necks and legs and the amount of feathering on their heads, and they can also be distinguished by slight differences in the texture and pores of their eggshells. Ostriches have the largest eyeballs of all the land animals: they are about the size of tennis balls, and take up so much space in their relatively small heads that they leave little room for the brain. Until recently the ostrich ranged throughout Africa, excluding the rain forests, but it has been exterminated in many regions. Domesticated ostriches that escaped or were released now occur ferally in North Africa, South Africa, and Australia. Four races or subspecies of the ostrich are recognized by most authorities.

North African Ostrich (*Struthio camelus camelus*)

This ostrich has a red neck and bare frontal head shield; it is the largest race or subspecies and therefore the largest bird in the world. When standing erect with its head held high it is 9 feet (2.74 m) tall, and the heaviest known specimen weighed 340 pounds (154 kg). Males have black-and-white feathering and the females are grayish. It originally had a very wide range, which included the whole of North Africa from the Mediterranean coast south into the wooded savannah bordering the central African rain forests, and from the Atlantic coast of Mauretania east to the Sudan and Ethiopia. It is now extinct in northern Africa for several hundred miles south of the Mediterranean Sea as a result of overhunting for meat and sport, and in the northeast corner of its current range it barely survives in southeastern Egypt bordering the Red Sea. It is now also rare throughout the rest of its natural range, but flocks of semi-feral and domesticated birds occur in some areas. Its rarity has been recognized by its inclusion in Appendix I of CITES, the international convention that controls trade in endangered species. The Ethiopian population of ostriches diminished in the late 1970s and 1980s when thousands of eggs were exported annually through the port of Djibouti, and many were also sold in the country. These eggs were obviously all collected from the wild as Ethiopia had no ostrich farms. This race of the ostrich was introduced in 1994 into the Mahazat As Sayd Protected Area in western Saudi Arabia, to replace the extinct Syrian ostrich, and others currently kept at Hai Bar, an Israeli Reserve north of Eilat, will eventually be released in the Negev Desert.

Somali Ostrich (*Struthio camelus molybdophanes*)

This race of the ostrich has a dull bluish-gray neck and thighs, which brighten in the breeding season. It has a bare head and whitish neck feathers, and while the male is typically black and white in adult plumage, the hens are duller and browner than the grayish females of the other races. It has a small range compared to the other races, being restricted to Somalia, eastern and southern Ethiopia, southeastern Sudan, and throughout Kenya, where in the south it overlaps the range of the Masai ostrich and hybridizes with it. The Somali ostrich survives in the harsh Ogaden region of desert and salt pans of eastern Ethiopia, which is claimed by neighboring Somalia and inhabited by nomadic tribesmen and their sheep and goats. Unlike the other ostriches, this race prefers bush and scrub country to the open plains, and usually lives solitarily or in pairs rather than in small flocks.

Masai Ostrich (*Struthio camelus masaiicus*)

The Masai ostrich has a pinkish-red neck like the North African race, and has small feathers on top of its head. It is the ostrich of East Africa, its range beginning in Kenya and extending south through Tanzania, Zambia, and northern Mozambique to the Zambezi River. Throughout this range it is still well respresented in undeveloped areas and in protected areas such as the national parks and game reserves.

South African Ostrich (*Struthio camelus australis*)

This race has a blue neck like the Somali ostrich but differs in having small feathers on its head. It originally occurred south of the Zambezi and Cunene rivers to the southern tip of Cape Province, but has been virtually exterminated in the wild except in the Kalahari and Namib deserts; its only safe havens are the national parks and other protected areas.

These long accepted subspecies of the ostrich were recently challenged by Sibley and Monroe in their *Distribution and Taxonomy of Birds of the World*. They recognize only two races—the African ostrich (*Struthio c. camelus*), occurring throughout Africa and originally into the Middle East, and the Somali ostrich (*S. camelus somalia*) of Somalia, Ethiopia, and Sudan, with both races occurring in northern Kenya.

Cassowaries

Cassowaries are elusive birds of the humid rain forest of the Australasian zoogeographic region, where they live in pairs or small groups in dense undergrowth from sea level to 10,000 feet (3,050 m) in mountain forest. They are the second-heaviest birds in the world after the ostrich, reaching a weight of 125 pounds (56 kg), but they are not as tall as the emu, standing 3 feet, 3 inches (1 m) to the top of their backs and almost 5 feet (1.52 m) to the top of their heads. There are three species of cassowary.

Southern, Common, or Double-wattled Cassowary (*Casuarius casuarius*)

This is the most well-known cassowary as it is the one most frequently available for zoos and bird gardens. It was first imported into Europe by the Dutch late in the sixteenth century, when it was called the Ceram cassowary. It occurs in Australia, New Guinea, and its neighboring islands, including New Britain, Ceram, Aru, Jobi, Japen, Schouten, and Waigeo. However, some of these islands may not be part of the species' original natural range, for the cassowaries of Japen, Ceram, and Aru are all believed to be imports from the mainland, involved years ago in "interisland" trading. In New Guinea's Vogelkop Peninsula and on Waigeo the cassowary's range extends almost to the equator.

Many races of this bird have been described but their range and biology is little known, with the exception of the Australian cassowary (*Casuarius c. johnsonii*). (See the color insert.) It is now restricted to the rain forest of northeastern Queensland, where it has a very patchy distribution. Externally, males and females are alike; their plumage, bills, faces, and tall standing casque are black and the rest of the head is

Australian Cassowary *The only ratite with a casque, this cassowary is mainly a fruit eater and is the second-heaviest bird in the world after the ostrich. It is now rare in the forests of northeastern Australia, unlike the other races of the species, which live in New Guinea, and which are still quite plentiful.*
Photo: Courtesy Ralf Schmode

bright blue with a red or orange streak at the back of the neck and a dark violet-blue lower neck. Two long, red wattles hang from the front of the neck and the powerful legs are greenish-gray. Its population is estimated to be no more than 3,000 birds, due to logging, which has reduced its favorite fruit trees (although replanting programs are now underway), and to roadkill on the roads that pass through its habitat. Cyclone Winifred destroyed rain forest in the vicinity of Innisfail in February 1986, and the local people fed the cassowaries that were forced outside their devastated habitat. Despite the cassowary's small territorial needs, its habitat has been so reduced that some young birds are unable to establish territories. The Queensland Fauna Conservation Act of 1974 listed it as a permanently protected species, so hunting is prohibited and a permit is needed to confine the species.

Northern or Single-wattled Cassowary (*Casuarius unappendiculatus*)

This cassowary is a bird of the lowland rain forest, from sea level to about 2,200 feet (670 m) in northern New Guinea, and also on the islands of Batanta, Salawati, and Japen. On average it is slightly larger than the previous species, and is therefore considered the largest cassowary. Its casque is also larger, and is sometimes more flattened than upright. It has a bare yellow-and-green neck with blue skin folds, one in the middle of the neck and two at the base of the beak, while its shoulders are reddish or golden. Hunting is the main threat to this species, especially in areas where logging and mining have provided access to its habitat.

Dwarf or Bennetts Cassowary (*Casuarius bennetti*)

Bennetts cassowary is the smallest species, although hardly a dwarf, as it reaches a height of 32 inches (80 cm) to the top of its back. It has a bare blue throat and neck, the back of its head is greenish, and it has a red stripe at the sides of its neck. This species has the most unusual black helmet or casque, which is small and flattened at the front and back, appearing almost triangular in shape. It is the cassowary of the mountains, a common bird of the high forests to an altitude of 10,000 feet (3,050 m) in southeastern New Guinea and the western Vogelkop Peninsula, and occurs also on the islands of New Britain, Ceram, and Japen.

Emus (*Dromaius novaehollandiae*)

The second-tallest bird after the ostrich, the emu reaches a height of 6 feet (1.8 m) to the top of its head and 4 feet (1.2 m) to its back. It is therefore taller than the largest cassowary, but not as heavy, averaging about 85 pounds (35.5 kg) when adult. The emus and cassowaries belong to separate but closely related families and are both placed in the order *Casuariiformes*. The emu has no crest and its face and throat are pale grayish-blue and are sparsely feathered. Its lower neck and thighs are thickly feathered, and like the body plumage are dark grayish-brown, becoming whitish at the base of the neck as the bird ages. The bill is black, the legs grayish-brown, and the eyes reddish-brown. Externally the sexes are alike.

The emu is distributed across the whole of mainland Australia west of the Great Dividing Range, except for the tropical forests of the North. It also occurred originally on Kangaroo and King Islands and in the state of Tasmania, but has long been extinct on all three islands. The emu is a bird of the deserts, grasslands, wooded savannah, and dry forest, where it lives in pairs or small parties. It is still very common in some areas, despite attempts to control its numbers, such as the 285,000 bounties that were paid in Western Australia between 1945 and 1960; and the bizarre "Emu War" of 1932–35, which was an embarrassing failure. During the campaign, a Lewis-gun detachment of the Royal Australian Artillery was sent to engage emus that were damaging the wheat fields. The gunners ambushed the birds at dams where they came to drink, and on one occasion fired point blank on 1,000 emus, but killed only a dozen before the gun jammed and the other emus dispersed. At another ambush site, only fifty birds out of a very large mob were killed, and the heavy military equipment was unsuitable for tracking individual birds when they scattered. Emus are sedentary where food and water is available, otherwise they are quite nomadic and range over great distances. In Western Australia banded birds travelled 300 miles (842 km) in nine months. Although several races of the emu have been described, one in eastern Australia and one each in the northwest and southwest, it is now more generally accepted that the species is monotypic.

Rheas

Rheas are the large flightless birds of South America, fourth in size to the ostrich, cassowary, and emu, and the only ratite to have evolved within the natural range of its ancestors, the tinamous. They are the most colonial of the grassland cursorial birds, roaming in flocks of up to thirty individuals outside the nesting season, and separating into smaller groups for breeding. They are named after Rhea, the sister and wife of the Greek mythological figure Cronus, youngest of the twelve Titans. Rheas are not closely related to the ostriches but are examples of convergent or parallel evolution, having evolved similar characteristics and habits as a result of living in the same kind of environment and experiencing similar pressures. Males are very aggressive birds when escorting their chicks, and are willing to tackle even cattle and vehicles if they believe them to be a threat.

Common or Greater Rhea (*Rhea americana*)

This is a bird of the lowlands, ranging from the Rio Negro in central Argentina (the northern boundary of Patagonia) north through the desert scrub, wooded savannah, and grasslands of the pampas and Gran Chaco into the scrub savannahs of northeastern Brazil almost to the Atlantic coast. Adult greater rheas have grayish wings and upper necks, darker feathers on the lower neck and back, and buffy-white feathers on their thighs and abdomens. White individuals are often seen. Males, which are one-third larger than the females, weigh about 55 pounds (25 kg), and are 5 feet (1.5 m) to the top of their heads. Their incubation period is thirty-five–forty days and the chicks are gray with dark stripes. Several races of the common rhea are recognized, the southern-most being Rothschild's rhea (*Rhea americana albescens*), which occurs in Argentina on the north bank of the Rio Negro. Others are *R.a. intermedia* of Uruguay; *R.a. americana*, which lives in central and northeastern Brazil; and *R.a. araneiceps* of the Pantanal region of southern Brazil and Paraguay.

Lesser Rhea (*Pterocnemia pennata*)

A slightly smaller bird than the greater rhea, this species has golden grayish-brown plumage with white flecks. It is very much a high altitude bird, living on the elevated, windswept scrub desert and short grass steppe of the Andean slopes, from the tip of South America to the Rio Negro in central Argentina, then north along the eastern slopes of the Andes to the grasslands of Peru's high altiplano. It has denser feathers than the common rhea due to the harsh nature of its habitat. The environment has also influenced its growth rate and the need to be able to withstand its first winter, so it matures very quickly, reaching full size at the age of four months and attaining adult plumage at twelve months. The lesser rhea is unafraid of water and has been seen wading into chest-deep water and then swimming with its wings held high over its back. The female is mute but the male has a booming call. Two races of the lesser rhea are currently recognized: the Puna rhea (*P. p. tarapacensis*), which occurs in high grasslands or altiplano at altitudes between 9,000 feet (2,750 m) and 12,000 feet (3,650 m) in southeastern Peru, northwestern Argentina, northern Chile, and western Bolivia; and Darwins rhea (*P. p. pennata*), a bird of the grassland and scrub from the Chilean province of Aisen south to the Straits of Magellan, which has been introduced onto Tierra del Fuego. Neither of their ranges overlaps that of the common rhea.

Lesser Rhea *The mountain rhea, a bird of the southern Andean slopes and the high grasslands of the altiplano— the high plateaus of Peru and Bolivia. It has also been introduced onto the island of Tierra del Fuego.*
Photo: Clive Roots

■ THE KIWIS OR SMALL RATITES

First sighted in 1642 by the Dutch navigator Abel Tasman, the three main islands and many smaller ones that comprise New Zealand lie almost 1,000 miles (1,600 km) across the Tasman Sea from Australia. The first Europeans to land there were surprised by the lack of mammals and the strange bird life, especially the number of flightless species. Yet the birds they saw were only a remnant of their former great variety, gone since the arrival of the Polynesians several centuries earlier. The islands were such a paradise for birds that they originally harbored not only the largest number of ratites the world has ever known, but also several other species that had lost their flight. When the ancestors of the Maoris arrived, the flightless birds included many moas, kiwis, rails, adzebills, a large parrot, a songbird, and a giant goose (*Cnemiornis*), which stood just over 3 feet (92 cm) tall. Only the kiwis, the parrot, and two species of rails survived, but even the remaining wildlife of New Zealand is so unusual that the islands are considered a subregion of the Australasian zoogeographic region or Notogaea, which is a very distinctive faunal zone.

Several factors influenced the development of New Zealand's modern fauna. Its early separation from the host land mass of Gondwanaland was undoubtedly the most important, for it occurred before the development of the mammals, so its animal life did not have to contend with the evolving mammalian predators. New Zealand's complete isolation since then, and its relatively small size, also influenced the evolution of its unique birds. The ancestors of its cursorial birds, which were on board as New Zealand drifted slowly across the South Pacific Ocean after breaking away from South America, evolved into many species, from the giant moas to the midget kiwis. Other land birds managed to reach New Zealand on the wind, and the paradise they found—the absence of predators, the mild climate, and the lack of competition for food—gave rise to other flightless forms that still survive there, and to several others which have almost lost their flight. Bats were the only land mammals to reach New Zealand by their own efforts, and the islands, which were later to become famous for their many flightless birds, had only two species of native mammals, both of which could fly.

New Zealand's relatively small size and its isolation were also critical factors in the evolution of its birds. They could not migrate to more favorable areas when their existence was threatened by climatic changes, as they could in North America when the polar ice cap covered almost half the continent; and the island's fauna consequently suffered severely during the Ice Ages. Climate change was so influential that the now extinct moas, the kiwis and a few other birds, the primitive tuatara lizard and some small frogs of the genus *Leiopelma* are believed to be the only survivors of New Zealand's original inhabitants when it broke away from Gondwanaland.

Kiwis are the only survivors of the largest concentration of ratites ever known, and together with the moas they developed the greatest wing reduction of all the cursorial birds. Their wings are so reduced they are not obvious externally, and the lack of functional wings, pectoral muscles, and tail has resulted in their distinctive pear-shaped body, so their scientific name or genus *Apteryx*, which means "bird without wings," is therefore quite appropriate. Unfortunately, the term "apterous," which means having no wings or wing-like extensions (from the Greek *apteros* meaning wingless), has become synonymous with "flightless" and "flightless birds." However, with few exceptions (the emu and cassowary, for example, which have rudimentary wings) the flightless birds have wings (often quite well-developed ones) which despite their size are useless for flight due to the deterioration of their flight muscles, and in some cases the great increase in the bird's body weight.

The kiwi's natural distribution was New Zealand's three main islands, plus D'Urville Island just off the coast of South Island, and possibly Little Barrier Island. They are not known to have occurred on the many other offshore islands, until recently at least, for they have been relocated to several that are predator free. Although kiwis were not harrassed by mammalian predators until recently, they were not entirely free from predation during their evolution. Subfossil bones show that large eagles once lived in New Zealand, but the weka rail, which evolved alongside them, is their only surviving natural predator, with a great liking for kiwi eggs and chicks. Kiwis are the most distinctive of New Zealand's birds, but like most nocturnal creatures are seldom seen. Originally forest dwellers, occurring in the native northern beech, kauri, and podocarp forests, they were forced into scrub, grassland, and exotic tree plantations as their habitat was reduced. Their nocturnal and secretive habits and need to hide from the now extinct diurnal raptors restricted their size development, unlike the other ratites, and the largest species only reaches a weight of 6 pounds (2.7 kg). Female kiwis are much larger than males, sometimes double their weight, and with correspondingly larger bill and legs. They have developed the typical powerful legs and sturdy feet of the ratites, which they use for digging, fighting, and running fast through the bush. Their leg muscles account for 30 percent of their body weight, and their legs are set so far apart that they run with a rolling gait. Their body temperature of 100°F (38°C) is about 3.6°F (2°C) lower than most birds.

Kiwis are somber birds, clad in blacks, grays, and browns, and streaked or mottled with paler shades. Although albinism is a disadvantage to most animals due to increased conspicuousness and predation, albino kiwis occur quite often and are at no greater risk because New Zealand's alien predators hunt mainly by smell. The kiwis' unusual feathering, which makes them appear hairy, is due to the loose structure of the feathers, which have poorly developed vanes and lack the barbules which would interlock them and keep their shape. They have small eyes and poor sight and their most developed senses are hearing, touch, and smell. The long and very sensitive whiskers at the base of their bills are elongated feather shafts, and the long, narrow bill with large olfactory bulb at its base gives the kiwis the most developed sense of smell of all birds. The nostrils are at the top of the upper mandible near its tip, and a lot of sneezing and blowing is needed to keep them clear of debris when they probe the soil for food.

Soil composition and texture are important elements of the kiwi's environment. It must be moist and have a high humus content, to provide the conditions needed for earthworms and for digging and probing for them. Earthworms form the bulk of their diet, and there is no shortage of them in New Zealand, which is home to almost 200 species. They have a great capacity for worms—a captive chick ate up to 300 every night—and as it is difficult to keep up with the demand, captive birds are generally weaned onto tofu and strips of raw liver. To locate worms, the kiwi first taps the ground with its bill and then uses it like a crowbar to open up a funnel-shaped hole, pulling the worm out steadily to prevent it from breaking. Any serious resistance by the worm, which would tear it apart, is met with a patient wait until it relaxes. To drink, the kiwi puts its lower bill into the water and tilts its head back to let it run down its throat; as it must never submerge the top bill where its nostrils are situated. Kiwis are certainly not water lovers, for they rarely drink and never bathe, and obviously derive sufficient moisture from their diet.

Like the large ratites, the kiwis are also very long-lived, with a maximum life span of almost thirty-five years. They are quarrelsome, territorial birds, and until he recognizes its smell, a male may attack his own chick when it returns to the nest burrow after foraging. With their aggressive ways and kicking power they resemble miniature cassowaries. They fight among themselves by kicking, the males hissing like geese or making high-pitched screams, while the females make deep roaring or growling sounds; but these occasions are just noisy affairs and they seldom break an opponent's tough skin. When they attack other animals and people, they kick forwards like the large ratites and their sharp claws and spur-like hind toes can inflict deep cuts. Male kiwis are sexually mature at eighteen months, but females do not lay eggs until they are at least three years old. The mated pair dig a burrow or find a suitable hollow beneath the roots of a tree, in a rock fissure, or under a clump of tussock grass, in which to build their nest. A single egg is normal for the kiwis, but the brown kiwi and the little spotted kiwi occasionally lay two eggs. When two eggs are laid, there is an interval of three–four weeks between them, and they hatch at the same interval, unlike the grassland ratite's many eggs whose hatching is synchronized so the male can leave the nest with almost a full brood of chicks. The kiwi egg is the largest of all birds in relation to the hen's size, in the largest species comparing favorably with an emu's egg and weighing up to 1 pound (450 g) or almost 25 percent of the female's body weight. Yolk forms 60 percent of

North Island Brown Kiwi   *This species is still relatively common but localized in the northern two-thirds of North Island, and has been relocated to several offshore islands, including Kawau Island in Hauraki Gulf near Auckland, and Kapiti Island, northwest of Wellington.*
Photo: Courtesy Department of Conservation, New Zealand. Crown Copyright. Photographer: Rod Morris, 1982

the egg's contents, which is also unique, as this is double the yolk size of most birds' eggs. The energy required to produce such a large egg is calculated at 1,250 percent of the bird's daily metabolism—the breakdown and absorbtion of nutrients—thus requiring a tremendous investment of calories by the hen. It is not surprising therefore that the laying of such large eggs, which need so much energy to produce, is well spaced. Although all other female birds have just one ovary, the left one, the kiwis have two, which function alternately if more than one egg is laid.

The kiwi breeding season is very long, lasting from June to March, and it was believed until recently that only the male was involved in the incubation, like most of the other ratites. Recent studies have shown that this is true only of the North Island brown kiwi and the little spotted kiwi. In the rowi and great spotted kiwi both sexes share the incubation duties, and the tokoeka actually breeds in groups, with the bonded male and female and other helpers all sharing the incubation. Kiwi eggs take between seventy and eighty days to hatch, equalling some of the albatrosses as the longest incubation period of all birds, and there is even a record of eighty-four days for the brown kiwi. The incubating bird places a single egg longways between its legs and a second egg is kept higher up near the neck. While the male is incubating, he is fed by the hen and also uses stored body fat, and he covers the egg with leaves when he occasionally leaves the nest burrow. When it hatches, the kiwi chick does not make a series of small holes around the egg to push off the end as most chicks do, but breaks off small pieces to make a hole into which it pokes a claw and tears the shell apart. The newly hatched chick is covered with inch-long black feathers and its eyes are open, but its legs are weak at first and it cannot stand until its fourth day. Little more than half of the egg yolk's stored energy is used by the embryo during incubation, and the balance is utilized by the chick in its first ten days, although in that time it may still lose up to one-third of its body weight. When it ventures from the nest tunnel on its sixth day it is totally independent of its parents for food, but they supervise its activities and force it into the tunnel at the first sign of danger. At first the chick is active both day and night, and by the time it is two weeks old it can run very fast.

New Zealand's flightless birds were hunted heavily by the Maoris in the absence of mammalian prey, and kiwis were killed for food and for their feathers, which were used in chieftains' ceremonial cloaks. When European settlement began, the island's natural mammalian fauna of just two species of bats had already been augmented by several aliens, including the kiore or Maori rat and the Maori dog (which accompanied the Polynesian colonists), and the black rats and feral goats and pigs introduced by sealers, whalers, and the early navigators. The dog and pig and possibly the black rat had already made a direct impact on the kiwis when the Europeans arrived, and the situation then rapidly worsened. Sir Walter Buller records professional hunters in North Island killing hundreds of kiwis for their skins, and additional alien predators were soon colonizing New Zealand's pristine forests and tussock grasslands. Stoats, weasels, ferrets, feral cats, and even the arboreal and normally vegetarian brush-tailed possum, all invaded the kiwi's burrows and ate their eggs and chicks. The transformation of great areas of the country through settlement and farming, especially logging and the clearing of bush for farmland, plus the subsequent erosion, has considerably reduced the kiwi's habitat and numbers. Dogs—feral, abandoned, or just out for the night—have caused tremendous damage to kiwi populations: the worst incident involved an abandoned German Shepherd in Waitangi State Forest that killed 500 birds in 1987.

Until recently, kiwis did not reproduce readily in captivity, but they are now regularly bred in small numbers in several New Zealand zoos and at the National Wildlife Center at Mt. Bruce. In addition, eggs are also collected from the wild for artificial incubation. The young are reared until about eight months old, at which size they are less vulnerable to the attacks of introduced predators, and are then released in protected areas. Kiwis were originally classified according to their anatomy, but their taxonomy has recently been revised as a result of DNA studies and five species are now recognized. Many translocations have been made to protect their surviving populations.

The Species

North Island Brown Kiwi (*Apteryx mantelli*)

This kiwi varies considerably in color and albinism frequently occurs. Generally, its head and neck are blackish-gray and its upperparts are dark rufus streaked with black, but in some birds the black is predominant, while in others there is so little black streaking that the rufus feathers are more prominent. The underparts are pale grayish-brown, the bill is light horn color, and the feet are brown. Its feathers are harsh to the touch. The brown kiwi is still a fairly common but localized bird on North Island, from Wanganui and Tongariro north into Northland. It occurs also on Little Barrier Island, which had an original population, possibly natural, prior to others being translocated there earlier this century. The saving of many of New Zealand's endangered species has depended upon the availability of islands free of predators, and kiwis have also been introduced onto Kawau and Ponui Islands in Hauraki Gulf, Motorua Island in the Bay of Islands, and Kapiti Island, northwest of Wellington, where the introduced possums were exterminated recently in a very intensive campaign. Ten North Island brown kiwis released onto predator-free Motukawanui Island in Matauri Bay in 1995 had increased to fifty birds by 2004. They were part of the program called Operation Nest Egg, which involved the removal of eggs and chicks from the nests of wild birds, and their being raised in captivity until they were able to fend for themselves.

Rowi (*Apteryx rowii*)

Rowi is the Maori name for this kiwi, which is restricted to the South Okarito Forest in Westland National Park. Until recently it was considered a race of the South Island brown kiwi and was called the Okarito brown kiwi, but was confirmed by genetic analysis in 2003 to be a distinct species. It is distinguished from the brown kiwi by its soft feathering and slightly grayish plumage with white facial feathers. Its small population has suffered severely from stoat and possum predation despite a trapping campaign for both animals. Artificial incubation of its eggs and releasing the chicks after they are sufficiently mature to survive is now considered the only recourse for this bird, the rarest of the full species of kiwis. The rowi's population now numbers 250 birds, and its habitat, 27,000 acres (11,000 ha) between Okarito and the Waiho River, has been declared a national kiwi sanctuary, one of five in New Zealand.

Tokoeka or South Island Brown Kiwi (*Apteryx australis*)

The tokoeka is a larger bird than the brown kiwi of North Island and has very soft, grayish-brown plumage, streaked with black. Its bill is horn colored and it has grayish-white feet. It was the first kiwi to be seen in England, when a specimen was sent home by Captain Barclay of the ship *Providence* in 1813. With a total population of about 25,000 birds the tokoeka equals the numbers of the North Island brown kiwi. It is quite a communal bird that is frequently seen in family groups, unlike most kiwis, which are decidedly antisocial and rather reclusive. It ranges from sea level—even inhabiting coastal sand dunes—to the mountain tops, where it may spend the winter, burrowing into the snow for shelter. The size of its territory depends mainly upon the quality of its food supply, and in earthworm-rich tussock grassland may be only 12 acres (5 ha) per family group. There are two races of the tokoeka, the southern tokoeka and the Haast tokoeka. The southern tokoeka occurs on both Stewart Island and South Island, but its range has been considerably reduced on the latter and it now lives only in Fiordland, where there are believed to be about 5,000 birds. It is still a common bird on Stewart Island, where it is the only species of kiwi. Inhabiting mainly the southern half of the island, it is breeding well and has become quite abundant lately with a current population estimated at 20,000 birds, despite the presence of feral cats. Fortunately, there are no introduced stoats, weasels, or pigs on Stewart Island. The southern tokoeka is now also established on Kapiti Island, where fifty birds live in the more moist areas which they share with the more plentiful little spotted kiwi.

The Haast tokoeka lives only in a small area on South Island's west coast, north of Fiordland, where it is isolated from other kiwis, and numbers between 200 and 300 birds. It is a very hardy bird, and although most individuals live at the fertile bases of the mountain slopes in the Haast, Olivine, and Selborne ranges and their river valleys in south Westland, others range up the wooded mountain sides to the subalpine grasslands, where they may even stay the winter. They burrow into the snow for shelter and dig under it to find earthworms in the soil, which must still be unfrozen due to the snow insulation, or they would never get enough food. The Haast tokoeka is a small bird with a very plump body, rufus plumage, and a distinctly down-curved bill. This rarest race of kiwi is protected in a sanctuary of 28,000 acres (11,400 ha) 15 miles from Haast Township, where a large stoat-trapping campaign is underway. It has also already been included in the Bank of New Zealand's Nest Egg Program and from two eggs collected in the wild in 2003, a chick has been hatched and raised, which was the first ever artificial incubation and raising of this kiwi.

Great Spotted Kiwi (*Apteryx haasti*)

This is the largest of the living kiwis, the bird which the Maoris called roa roa, with hens weighing up to 6 pounds (2.7 kg), but the males are just half that weight.

Great Spotted Kiwi   *The largest kiwi, weighing up to 6 pounds (2.7 kg), this species lives on the western side of New Zealand's Southern Alps, from sea level to the alpine heights. At lower elevations it has been preyed upon heavily by introduced stoats and by feral dogs and pigs.*
Photo: Courtesy Department of Conservation, New Zealand. Crown Copyright. Photographer: Rod Morris, 1975

Its head and neck are dark grayish-brown, and its upperparts are fulvus, mottled, and banded with brownish-black, giving it a spotted appearance. Its gray underparts are also similarly mottled. The great spotted kiwi lives only on South Island, on the west side of the Southern Alps, where it is split into three distinct populations. The largest of these lives in northwest Nelson, especially in Kahurangi National Park, with another in the Paparoa Range and the third in the Southern Alps from Lake Sumner to Arthur's Pass. It is a very hardy bird, which is now most plentiful in wet and mossy subalpine areas, for in lowland and coastal forest it has been preyed upon heavily by dogs, pigs, and stoats and has declined in many places. Fortunately, these predators rarely venture up into the higher altitudes. As its present distribution is mostly in protected areas, however, it is at least not threatened by habitat destruction. Several were released in 2004 into the beech forest on the shores of Lake Rotoiti in Nelson Lakes National Park, the first species to be reintroduced there after an intensive campaign to remove all alien predators, and they are the first kiwis to live there for eighty years. This species is believed to have always been restricted to South Island, despite the land bridge between it and North Island, which existed until the end of the last Ice Age. In 2003, the great spotted kiwi's current population was estimated to be about 17,000 birds.

Little Spotted Kiwi (*Apteryx oweni*)

The little spotted kiwi is the smallest species, with females weighing up to 4¾ pounds (2.2 kg) and males just 2¾ pounds (1.5 kg). It has yellowish-gray feathers on its head and neck, and its body is similar but lighter and is banded and mottled with blackish-brown, which produces the spotted effect. The bill is flesh-colored and the feet are off-white. This species was apparently a common bird in the middle of the last century, but it is now one of the rarest kiwis. Its habitat is woodland and tussock grassland, originally on the western side of South Island from northeastern Marlborough to southern Fiordland and on D'Urville Island. It also occurred on North Island, as a few birds were discovered in 1875 on Mt. Hector near Wellington, just below the snow line in mossy subalpine bush, but this colony has long since vanished. It has also disappeared from D'Urville Island and surveys in 1981 found no trace of the bird in its original South Island habitat. Fortunately, it was introduced to Kaptiti Island (12 square miles/19.3 square kilometers) early this century, where the population now numbers 1,200 birds despite heavy egg predation by weka rails and rats, which have now been eliminated from the island by a massive eradication campaign. It was also more recently introduced to Red Mercury, Hen, and Tiritiri Matangi Islands, which each now have small but thriving populations. Believed extinct in its natural range for several years, a little spotted kiwi was caught in Westland National Park in 1992, and was released on Mana Island because it is genetically different from the others of its race on neighboring Kapiti Island.

Notes

1. The grasslands of central Argentina.
2. The treed plains south of the equator in Brazil.
3. The high Andean grasslands of Peru, Bolivia, and Chile.
4. The last and continuing portion of the Quaternary period, beginning about 10,000 years ago.
5. In addition to the ratites many other birds, especially numerous pigeons, woodpeckers, and parrots, have also lost the oil gland.

3 Blown to Paradise

The water that covers almost three quarters of the earth's surface has had a profound effect upon evolution. Although the vast expanses of the oceans acted as an insurmountable barrier and limited the natural spread of most plants and animals, remote islands offered great opportunities for the development of unique species, if they could get there. Despite the tremendous difficulties involved in reaching distant islands, and of overcoming the problems associated with settlement and the establishment of viable communities, plants and animals took advantage of these opportunities, and those that succeeded evolved into some of the world's most distinctive forms of life.

Most islands are continental and originated from their neighboring land masses. They include offshore islands like the British Isles, Vancouver Island, and Newfoundland, plus more isolated ones such as South Georgia, Madagascar, New Zealand, and even far-flung Kerguelen Island in the South Indian Ocean. They all separated from their associated land masses long ago, either by slowly drifting away or as a result of sea-level changes; and their wildlife is descended from the species that were on board at the time of separation, plus the immigrants that reached them later.

Oceanic islands are of greater evolutionary interest than most continental ones, because they arose dramatically from the seabed as a result of volcanic action, and prior to their upheaval were never connected to the continents above water. Many are thousands of miles from the nearest land mass and surrounded by deep water, which is the greatest barrier to the movement of nonflying land animals. But even though they are the world's most remote habitats, organisms managed to colonize them, although their life forms are limited because of the difficulties of reaching such a distant speck of land in a vast sea. Their uniqueness stems from the immigrant organisms' responses to their new environment, and their subsequent endemism—the evolution of forms found only on their particular island, which differ from their ancestors.

Uninhabitable when they first erupt from the sea, oceanic islands are colonized very slowly as crumbled lava and windblown dust collect in crevices. Germinating seeds can then survive and a plant community becomes established. This is the first priority, for without vegetation few colonizing forms of animal life can exist. With the exception of wide-ranging seabirds, oceanic islands were colonized mainly by luck—by a coconut washing up on the beach, a windblown seed lodging between rocks, or a gravid (pregnant) lizard still clinging to driftwood after weeks at sea. But oceanic island colonization was a very slow process and it has been estimated, for example, that the oldest of the Hawaiian Islands erupted above sea level about 10 million years ago, so their 400 basic species of plants have colonized them at the rate of only about one every 2,500 years.

Plants reached remote islands by wind dispersal, on ocean currents, or attached to seabirds. Seeds and spores, especially those with hooks and barbs to aid their dispersal, may be carried on bird's feathers or feet; and several have actually been found in the feathers of the yellow-nosed albatross (*Thalassarche chlorohrynchos*), which ranges across the southern Atlantic Ocean and nests on islands of the mid-oceanic Tristan da Cunha group. Bouyant seeds with protective husks have been carried thousands of miles on ocean currents and survived to sprout on a distant beach. Winged invertebrates were mostly carried to the new islands on the wind, although not too successfully, for insect life on remote islands is still rather sparse. Where insects are concerned, colonization has been limited by the uncertainty of being blown there in the first place, the inability to overcome such problems as the lack of standing fresh water for their larval development, or simply the fact that insects arrived before their food plant or a suitable alternative reached the island. Reptiles and amphibians are also poorly represented on most oceanic islands because they were totally dependant on oceanic currents and driftwood to carry them there.

Animals independent of their new island for food are the first to colonize it successfully. Sea lions and fur seals rest and calve on the beaches, and seabirds nest in their millions in the absence of predators. The island may be just a barren rock but the marine mammals and birds have direct access to the rich seas, which support their large numbers. Their detritus adds to the buildup of soil, improving the chances of plant growth and the island's eventual colonization by other organisms. Windblown land birds arriving on remote newly formed islands had no access to these rich sources of seafood and had to scavenge on the beaches; with few exceptions their colonizing efforts took many years.

An interesting aspect of island colonization is that windborne and waterborne settlers often arrived on their new island from different sources. Ascension Island, in the middle of the Atlantic Ocean, received plants from Africa because it lies in the path of the Benguela Current, which flows north past the continent's southwest coast before being turned into mid-Atlantic by the bulge of West Africa. Yet the ancestors of the Ascension Island rail, alas now extinct, are believed to have been wind-assisted from South America. Similarly, Fiji's endemic iguanas are undoubtedly descended from green iguanas ocean-borne on driftwood across the Pacific Ocean from South America, whereas the ancestors of its rails island-hopped in the opposite direction from Southeast Asia.

Birds are the most plentiful vertebrates on remote islands, and some marine species nest there in the millions, but few oceanic islands have a wide range of resident land birds. Whereas invertebrates, amphibians, and reptiles have colonized many isolated islands by water, surviving the long crossings clinging to driftwood or a clump of matted vegetation, it is unlikely that any land bird has ever reached distant islands alive in this manner. Apart from the transportation of birds as food by human colonists, such as the Polynesians with their jungle fowl and possibly megapodes, the only natural way for birds to travel was by air, and this happened accidentally. It is well known that storms blowing off land, or across islands, carry birds out to sea, and after major gales many bird carcasses have been seen on the surface far from shore. The difficulty of locating a remote speck of land in a vast ocean in this haphazard fashion can be appreciated, and landfalls were obviously due to luck. The fact remains, however, that land birds did manage to reach the world's most remote oceanic islands, in tropical, temperate and even sub-Antarctic seas, and as a group the members of the family *Rallidae* were the most successful in reaching and colonizing them.

Like the early human colonists of the world's remote places, times were undoubtedly hard for birds arriving on young oceanic islands. Those coming in later years, after an island's colonization by seed and fruit-producing plants and invertebrates, and with no predators or competition for food, found paradise. Similarly, birds blown to distant continental islands like New Zealand—from the evolutionary point of view the most interesting islands on earth—and with established flora and fauna but no predators, also found paradise on earth. Rails and their relatives reached many of these remote islands and found them perfect for their way of life. In time their descendants produced more flightless forms than any other group of land birds, and after long isolation on remote islands they have evolved into several monotypic genera—unique birds with just a single species in each genus.

■ THE *RALLIDAE* FAMILY

The *Rallidae* is a very ancient group of birds, whose members live mainly in wetlands where there is shallow water and dense vegetation. The family is normally divided into three groups. The rails are secretive hen-like birds of the marshes; the gallinules or moorhens[1] are more aquatic in their habits and are seldom found far from open water; and the coots are even more aquatic, having lobed toes for improved swimming ability. In appearance and habits the gallinules and coots are similar, but for the purposes of this chapter the rails and gallinules are grouped together, and referred to as rails. Apart from convenience, this is because the habitat and habits of several of the now flightless insular descendants of the gallinules have changed considerably, and some are less aquatic than many of the rails. For example, the Gough Island moorhen lives in dense grassy vegetation, the Samoan moorhen occurs in upland forest, and the San Cristobal moorhen, if it still survives, occurs in dense forest on precipitous mountain slopes. The takahe, an overgrown gallinule, lives in the tussock grasslands of mountain valleys in New Zealand's South Island. The only truly aquatic flightless member of the family—the giant coot—is considered a waterbird and is therefore included in Chapter 6.

The rails have long toes, short and rounded wings, short tails, and laterally compressed bodies that make it easier for them to pass through thick vegetation. They are agile climbers, able to flutter and scramble into trees and bushes—even the nonfliers may nest 30 feet (9 m) above ground—and they swim well despite their lack of webbed feet. Many rails are nocturnal, most are very secretive, and with the exception of the totally vegetarian takahe, they are omnivorous. In fact, they eat virtually anything organic from seeds, fruit, and berries to invertebrates, birds' eggs and nestlings, lizards, and small mammals, and like the thrushes they hit snails on rock "anvils" to open them. Rails forage on the beaches, turning stones over to find crabs and molluscs; they eat seabirds' eggs and the food that seabirds regurgitate for their chicks, which is how some survived during their early colonizing days when nothing else was available. They can cause serious damage when they probe the soil for earthworms, and the large alien weka rail on Mac-Quarie Island, which was introduced as food for visiting whalers years ago, is considered a pest as it has eroded the topsoil in places with its probing for worms.

Although many insular rails are now flightless, their ancestors could all fly and most of the 135 species in the family *Rallidae* still fly. However, flight in rails is rather a contradiction. Some make long annual migrations and others are poor fliers, but without exception they all fly reluctantly. Many rails are summer visitors to the northern temperate regions. From Europe they fly back to Africa in the fall and the species that breed in northeastern Asia migrate south to India, the Malay Peninsula, and Indonesia to escape the harsh winters. Similarly, rails in the New World fly south from Canada to the neotropics for several months each year. However, even the continental species, which make long seasonal flights, are difficult to flush, as hunters well know, and with long legs dangling they fly a short distance before dropping back into the undergrowth. Such seasonal inactivity between long migratory journeys is obviously insufficient to compromise flight, but many rails live in tropical climates, do not migrate, and could therefore be expected to lose their flight because of their general reluctance to get airborne. Species such as the white-breasted waterhen—a common marsh bird in the Orient—the tropical American gray-necked woodrail, and the lesser gallinule of Africa all have typical rail habits but are still able to fly. There are two reasons for this: none live on islands and the risk of predation in their habitat is high.

Rails were probably not blown to sea because they were such poor fliers; such skulking, low-flying birds were surely less likely to be blown off land than high flyers during violent storms. The rails must have had a higher wind tolerance than most birds, but when they were blown away, the ancestors of the insular flightless rails were able to stay aloft long enough to reach the world's most remote oceanic islands, including those in the mid-Pacific and mid-Atlantic oceans, thousands of miles from the nearest mainland.

In the southern oceans there is a region of stormy seas, between 40° and 50° latitude, known as the roaring forties. The strong westerly winds which blow in this area of uninterrupted seas have been a major force in bird dispersal; their powerful winds carried birds to new homes and were responsible for many successful natural translocations. From the southeastern coast of South America winds blew purple gallinules (*Porphyrula martinica*) almost 2,500 miles (4,000 km) to islands in the mid–South Atlantic, and gave rise to the now extinct Tristan gallinule and the subspecies that still lives on neighboring Gough Island. The Auckland Islands lie right in the path of the roaring forties, which blow from the west most of the time, yet twenty forms of birds, including the now rare flightless rail, reached the islands from New Zealand, 250 miles to the north, and became established. The Tasmanian white-eye, which still flies, reached New Zealand in the middle of the last century after being blown across 1,000 miles (1,610 km) of open ocean with no land in between for island-hopping or staging, a remarkable journey for such a tiny, nonmigratory species, which is now New Zealand's commonest bird.

Wind-borne rails, marooned on mid-oceanic islands like those of the Tristan da Cunha group, had no hope of returning to the mainland and little chance of reaching islands beyond the others in their group even while still capable of flight, let alone once they had lost the ability. Others were more fortunate and island-hopped to extend their range, eventually evolving into new forms, and like the later Polynesian seafarers, rails colonized the islands arcing 7,500 miles (12,000 km) across the central and southern Pacific Ocean. The prevailing winds normally blow westward across the Pacific, but occasionally change direction, giving rise to violent cyclones that can sweep a small island clear of its vegetation and blow most airborne creatures out to sea. Island hoppers were obviously blown to other islands while they could still fly well enough to stay aloft and aid their dispersal, and flightlessness must have been well underway before birds were no longer at the mercy of high winds. Rails still able to fly when blown offshore were either lost at sea during the first lull in the wind, or by chance landed on islands hundreds of miles to the east. From Southeast Asia they migrated in this manner through the Caroline Islands, the Gilbert and Ellice Islands, Fiji, Samoa, the Society Islands, and the Tuamotu Archipelago, all the way to Easter Island. Although individually small and remote, most islands in these far-flung groups are only a few hundred miles apart, and for birds able to make much longer sea crossings, these short journeys must have been relatively easy. This did not just happen in the South Pacific Ocean, however. It is 250 miles (400 km) from Madagascar to the Aldabra Islands and 300 miles (483 km) from Australia to Lord Howe Island, both short hops for the rails, which managed the journey to both islands, settled there, and lost their wing power. The rails that were blown eastward and colonized new islands developed characteristics of their own, leaving others behind to continue evolving along their own lines. Further regression toward flightlessness decreased the risk of being blown away, and hastened the process in those left behind. From west to east across the Pacific Ocean several unique island forms evolved, including the Iwo Jima rail, Kittlitz's rail of Kusaie, the New Caledonia rail, Samoan rail, red-billed Tahiti rail, Henderson Island rail, and the Easter Island rail.

Crossing the sea successfully was only the first obstacle, for safe arrival on an island held no guarantees of individual survival or eventual colonization. The disoriented, exhausted, and hungry bird had to find food and water quickly. Those that arrived early in an island's development before even plants were established could not survive unless they were expert opportunists, which the rails certainly

Purple Gallinule *Long ago winds blew this common South American rail out to sea and deposited them on Tristan da Cunha Island and Gough Island in the mid-Atlantic, where they evolved into two endemic species, although the Tristan gallinule is now extinct. Evolution has so changed the Gough moorhens that they do not recognize purple gallinules recently blown to the island as potential breeding mates.*
Photo: Stefan Ekernas, Shutterstock.com

were. The now extinct Ascension Island rail survived on its barren home on seabirds' eggs and the food they regurgitated for their chicks, but few birds have such liberal eating habits, and many must have perished soon after their landfall.

When new arrivals found adequate food and water and no predators, colonization was still not assured. A gravid black rat gaining the shore of paradise down a ship's mooring line was virtually guaranteed the opportunity of founding a whole island population, but birds had to find a mate of the right sex and of breeding age. When marooning was such a random occurrence, the chance of a suitable mate arriving at the right time was obviously poor. Even then, one pair of birds may not have been able to colonize an island, because inbreeding may eventually have affected the population's viability. However, the lack of genetic variation due to the small number of founders in an enclosed population is not a disadvantage to some birds, at least under natural conditions. A classic case is that of the pair of ultra-marine lories—rare brush-tongued parakeets—introduced onto the island of Ua Huka in the Marquesas, which resulted fifty years later in an apparently healthy population of 250 birds. Rails are adaptable, opportunistic birds, whose reproductive habits aided their colonizing efforts. Prolific breeders, a pair can annually produce three broods of four or more chicks each, and to achieve this they drive their young away at an early age to find mates and establish their own territories. The absence of predators and a whole new island to colonize gave the rails unrivalled settlement opportunities, and it is entirely possible that some small endemic populations stemmed from just a single breeding pair.

When birds successsfully overcame the problems posed by their new environment, such as finding food, water, and mates, and in the absence of predators and competition from similar species, their new home was indeed utopia. The different environments produced changes in their habits, then in their size, color, and even feather structure, resulting in new forms endemic to their island. Naturally, immigration did not cease because a new species or race had evolved there. At some stage however, the changing insular birds became so distinct in their appearance and ways that breeding did not occur between them and the members of their ancestral species that later found their way to the island. The Gough Island moorhen, for example, which evolved from windblown South American purple gallinules, no longer recognizes as potential mates the gallinules that still occasionally reach the island. Eventually, the continued arrival of the same ancestral stock even gave rise to other distinct forms. In New Zealand, the takahe evolved long ago from the purple swamphen (a gallinule) that more recently also produced the pukeko (*Porphyrio porphyrio*), which still flies. New Zealand's other flightless rail, the weka, is believed to be descended from an early immigration of Australian banded rails (*Gallirallus philippensis*), of which the island's current population of the species (which stems from a later colonization), has changed sufficiently to be considered a subspecies, although they still fly.

Flightlessness was not solely the prerogative of the rails. Island life had the same effect upon other birds such as the pigeons, which gave rise to the now extinct dodo and the solitaires, and upon the now semiflightless Henderson Island fruit pigeon. Loss of flight was the major change to occur in all the flying animals, excluding the bats, which settled on remote islands. Able to find food without flying, and with no predators harrassing them, many species became flightless, and some even wingless, and this happened to numerous insects as well as birds. Like the rails of the Tristan da Cunha Islands, the descendants of several species of moths that were blown there have also lost their flight. There are wingless wasps and flightless flies on the Hawaiian Islands, and on the Galapagos Islands the giant painted grasshopper has given rise to the flightless short-winged grasshopper. Some insects, like some birds, were obviously more prone to being blown out to sea, or at least to surviving until they reached land, and any trend toward flightlessnesss eventually gave them an advantage over the winged forms when storms blew across their new homes.

The flightless land rails live only on islands; there are none on the continental landmasses, but although the rails' behavior may have predisposed them for flightlessness, this did not happen until they reached islands with ideal conditions. The most important criterion for the loss of flight was the level of predation, not necessarily its total absence. Rails lost the use of their wings on all the predator-free islands that they colonized, but the same also happened in Tasmania and New Guinea, which have some very opportunistic carnivores. In fact, the recent marsupial carnivores were on Tasmania when it became separated from mainland Australia. They were the thylacine or Tasmanian wolf, the Tasmanian devil, and two spotted native cats—the tiger quoll and the eastern quoll. All thrived there until the wolf's demise in the last century, but neither it nor the devil could compete with the dingo on the Australian mainland and both have been extinct there for some time. In Australia, the black-tailed native hen (*Gallinula ventralis*) still flies, yet in Tasmania it gave rise to the flightless Tasmanian native hen (*Gallinula mortierii*), despite the presence of the carnivores. It is still a common bird on the island, even with the additional threats from introduced rats, cats, and dogs.

There is obviously something about islands where rails are concerned. Once established on them evolution has virtually guaranteed that a flightless form will develop, predators or not. The island's size appears to play an important part in both stimulating flightlessness and allowing it to develop more quickly on small islands. This is not surprising, for if the basic need to fly was removed, loss of flight would be genetically promoted by inbreeding because of the small number of founders in the original stock, the small population which the island could support, and the absence of new blood. With the exception of New Guinea—the world's second-largest island—and the main islands of New Zealand, most of the twenty-three known living forms of flightless land rails, and the nineteen recently extinct ones, evolved on much smaller islands, several of them less than 20 square miles (32 square km) in area. Until recently, small tropical islands were indeed a paradise for the land rails.

The Species

Weka or Woodhen (*Gallirallus australis*)

The weka is a New Zealand rail the size of a small chicken, with adult males weighing just over 2 pounds (900 g). It is cosmopolitan in its choice of habitat, and lives in coastal swamps, scrubland, temperate rain forest, and tussock grassland well above the treeline. It is active both day and night, swims well like all rails and runs very fast, fluttering its wings as it does. Although flightless, the weka actually has large wings, but they are poorly developed and feeble, the quill feathers are soft and flexible and the long secondaries resemble a loose mantle, and of course its breast muscles have degenerated. It is believed to be descended from a very early invasion of the islands by the banded rail, a much smaller bird (only 10 inches/28 cm long), which also lives in New Zealand as a flying species that probably stems from a more recent colonization. The weka is an inquisitive bird that becomes tame very quickly and frequents picnic sites and campgrounds looking for handouts. It is fond of chicken eggs and taps a hole in the shell to siphon out the contents.

The weka is a prolific breeder, laying up to four clutches of eggs each year in a nest of woven grass. It also digs burrows with its bill, where it may shelter during the day and may even nest there. Each clutch contains three creamy-white eggs with scattered purple-and-white blotches, which hatch after an incubation period of twenty-six days. The parents share the incubation and chick-raising duties, and the young are driven away when only eight weeks old, so that nesting may begin again.

Weka   The flightless wekas of New Zealand are believed to have evolved from banded rails blown across the Tasman Sea from Australia long ago. They are aggressive birds that eat kiwi eggs and chicks, and are in turn preyed upon by introduced carnivores like the stoat and polecat, and feral pigs and dogs; which have exterminated wekas in many regions.
Photo: Courtesy Alan Liefting

The weka is an aggressive bird, especially when raising chicks, and although helpless against stoats and weasels, it is more than a match for rats, even acting as a "guardian" for smaller birds nesting nearby. If the chicks wander off and do not respond to their parents' calls, they are picked up by the nape of the neck and carried back to the nest, a frequent mammalian trait but very uncommon behavior in birds.

There are three insular races of the weka rail. The North Island weka (*Gallirallus australis greyi*) was exterminated in many areas by feral cats and dogs and introduced weasels and stoats. It is now making a comeback due in large part to the reintroduction of captive-bred birds. However, the presence of alien carnivores is still a problem and the reintroduction program suffered a setback in 1995 when most of the birds released near Waihi were killed by ferrets. Safe island sites were therefore considered for future releases, and birds have already been introduced to Kapiti Island. The buff or eastern weka (*Gallirallus australis hectori*) originally occurred on the eastern side of South Island, but was extinct there by 1924, possibly as a result of introduced bird diseases. Fortunately, it had been introduced earlier onto Chatham Island, where it is thriving. However, it has since been determined by DNA testing to be just a color phase of the black weka (*Gallirallus a. australis*), which is the race occurring in south Westland and western Otago in South Island where it is still a fairly plentiful species, and it has also been introduced onto Kapiti Island. The "buff weka" of Chatham Island now has a great opportunity to evolve along its own lines. The third race, the Stewart Island weka (*Gallirallus australis scotti*), is still well represented on Stewart Island and was introduced onto MacQuarie Island where it is also thriving.

Okinawa Rail (*Rallus okinawa*)

Okinawa is the largest island in the Ryukyu Archipelago, which stretches over 620 miles (1,000 km) in the East China Sea from the southern end of Japan (Kyushu) to Taipei, beginning about 370 miles (600 km) from Kyushu. A subtropical island of 450 square miles (725 square km), Okinawa was heavily forested, but most of the southern forest has been cleared for agriculture, especially sugar cane plantations. It is most famous for one of World War II's bloodiest campaigns. Its rail is endemic, and lives in dense undergrowth near water in evergreen forests in the mountains of the Yanbaru region in northern Okinawa. It was only discovered in 1978, and officially named in 1981, so little is known of its biology, but it has short wings and is definitely flightless, fluttering up sloping branches to roost above ground. A small bird, weighing just 1½ ounces (42 g), the Okinawa rail is brown with pale barring on the breast and stripes on the head and neck, and has heavy, orange-colored legs and feet. It was believed extinct until recently, due to road and dam construction, habitat destruction, and predation by cats and the mongooses that were introduced to control the venomous pit viper called the habu. However, ornithologist M. A. Brazil induced over 150 responses to his recorded rail calls as he travelled through the north Okinawa forest on five consecutive nights, and the population is believed to number about 1,000 birds. Current conservation measures include the possibility of building a very large wire mesh enclosure for breeding, the artificial raising of chicks, and the reduction of the alien mongooses—which has been tried before with little success.

Henderson Island Rail (*Porzana atra*)

Henderson Island is a small raised coral atoll of about 14 square miles (22 square km) lying 107 miles (172 km) east of Pitcairn Island in the southeastern Pacific

Ocean. It is edged with steep cliffs of bare limestone, and is mainly densely vegetated with scrub and low trees such as screw pine (*Pandanus tectorius*) and sandalwood (*Santalum hendersonense*). It is almost surrounded by a fringing reef up to 250 feet (75 m) wide, lying about 60 feet (18 m) offshore, with two narrow access channels for boats. With its ecosytems practically intact, it is considered the best remaining example in the world of a raised coral ecosystem, but at its highest point it is only just over 100 feet (30 m) above sea level and is therefore vulnerable to storm damage. Its isolation has encouraged the evolution of a unique fauna and flora, with ten endemic species of plants and four birds, one being a totally flightless rail, another a fruit dove which has almost lost its flight. It also supports a large breeding population of seabirds—up to 80,000 pairs—of petrels, tropic birds, frigate birds, and boobies, and is important as a wintering place for northern migrants. Since the extinction of the Easter Island rail the resident rail of Henderson Island is the easternmost flightless bird in the South Pacific Ocean.

A World Heritage site, Henderson Island has been uninhabited since the Polynesians vacated it at the end of the fifteenth century, and the only signs of previous habitation are the coconut groves and fruit trees that grow near the main landing beach, plus the inevitable Polynesian rat, which accompanied the colonizers to many islands across the Pacific Ocean. Fortunately, the goats and pigs that were also released there did not survive. Its isolation provides the island with some protection and it is now visited only by passing yatchsmen, occasional cruise ships, and the Pitcairners who collect firewood, fruit, and coconuts. It is an arid island, its only surface water being the rainwater that collects in small pools, but this is usually quite brackish due to salt spray and evaporation. Its rail, a small, tame, and friendly bird, has blackish feathering and reddish-orange feet and, as it cannot fly, it is restricted to searching the forest floor for insects and small molluscs. It is believed to be descended from the spotless crake, a widespread species on South Sea islands and in Australasia. The Polynesian rat, originally not considered a threat to ground-nesting birds, has preyed heavily upon the chicks of the dark herald petrel, an endangered species that nests on the island, and is therefore believed to have also affected the rail's numbers. Despite the rats, and the large crabs which are known to eat rail chicks, the Henderson Island rail population is considered to be stable at about 4,700 pairs.

Aldabra Island Rail (*Canirallus cuvieri aldabranus*)

Aldabra is not an individual island, but the name of a remote atoll comprising four main islands—Grande Terre, Malabar, Picard, and Polymnie Islands—positioned around a shallow lagoon which contains other islets, situated 250 miles (400 km) north of Madagascar, and 680 miles (1,100 km) southwest of Mahe in the Seychelles. It is the world's largest raised coral atoll, the islands being the coral-covered tips of a volcanic seamount, fringed by a reef, rising to only about 30 feet (9 m) above sea level and in total covering an area of 96 square miles (155 square km). The vegetation includes mangroves and casuarina along the shoreline and low mixed scrub with open inland areas of grass and flowering plants (of which

40 species are endemic). It is more famous for its giant tortoises than its rail, and is also the home of a large number of breeding seabirds, especially terns and tropic birds.

Aldabra is believed to have been first visited by Portuguese navigators early in the sixteenth century, but settlement was discouraged due to its arid nature, lack of fresh water on all except Grand Terre, and its distance from the main sea lanes across the Indian Ocean. Settlement of the neighboring Seychelles in the eighteenth century resulted in more frequent visits to Aldabra to collect giant tortoises for food, and it was during this time that both black rats and domestic cats were introduced, although the atoll has since remained relatively free of human disturbance.

The Aldabra rail, a medium-sized species about 11 inches (28 cm) long, is a pretty, slender bird with pale maroon head and breast and a white throat; and is restricted to the islands of the atoll. It is a race of the white-throated rail (*Canirallus cuvieri*) of nearby Madagascar that also gave rise to the now extinct flightless rail of Assumption Island. Historically, the rail occurred on several neighboring islands in addition to those of the Aldabra Atoll, but a century ago survived only on Polymnie and Malabar Islands. Currently, most live on Malabar Island, where the population is said to be about 1,000 birds, with a few on Polymnie Island and on the islets Ile

Tasmanian Native Hen *This rail is a descendant of the smaller and still flighted black-tailed native hen of mainland Australia. Isolated on Tasmania for many years, it became flightless despite the presence of major carnivores including the Tasmanian devil and, until recently, the Tasmanian wolf or thylacine.*
Photo: Courtesy Malcolm Laird

Michel and Ile Aux Cedres. When Picard Island was recently declared cat-free, several rails from Malabar were transferred there and are thriving.

The Aldabra rail has a close association with the islands' giant tortoises, picking parasites off their carapaces and eating the invertebrates in their dung. Like many birds on remote islands it is tame and inquisitive. It has shortened wings which it rarely opens and, together with the pronounced reduction of its pectoral muscles, is incapable of flight. The major threat to its continued survival comes from the introduced cats and rats, and although an adult bird was seen killing a black rat, its eggs and chicks are vulnerable. The Aldabra rail is the last flightless bird of the western Indian Ocean Islands.

Tasmanian Native Hen (*Tribonyx mortieri*)

The large endemic flightless rail of Tasmania is probably the most secure of all the rails that can no longer fly; it is common and frequently seen in open grassy areas. It is a dull greenish-brown bird, descended from the black-tailed native hen of Australia; and although it is uncertain how long it has been isolated on Tasmania, it has certainly been long enough for it to lose its flight and increase in size to 17 inches (43 cm) long, while its flying ancestor in Australia is just 13 inches (33 cm) in length. Males weigh up to 2½ pounds (1.1 kg), while the similar plumaged females are a little smaller. The most surprising aspect of this case of island colonization and loss of flight, however, is that it occurred in the presence of all the largest recent marsupial carnivores—two species of native cats or dasyures, the Tasmanian devil, and until recently the largest of all, the thylacine or Tasmanian wolf. The native hen survived, lost its flight, and is still common despite the natural predators and the arrival in the last two centuries of European settlers with their cats, dogs, and rats. It has also survived official "vermin" extermination programs based on its supposed destruction of crops, and between 1955 and 1958 thousands were killed.

The Tasmanian native hen feeds in the open but it is a very timid bird with well-developed flight reflexes; it runs back swiftly to the safety of thick cover when approached. It has been clocked at 28 mph (45 kph), and while running fast uses its short wings for balance. The native hen's prolific breeding habits have also aided its survival, for it lays up to nine eggs in a clutch, at least twice each season, in the Austral spring and summer. Also, as males usually predominate in a population, a female will mate with two possibly related males—a practice called polyandry which is rare in birds—so there is not the same selective pressure to fight for the hen's favor. It has been introduced by the Parks and Wildlife Service to Maria Island National Park off Tasmania's east coast.

New Britain Rail (*Gallirallus insignis*)

This rail is endemic to New Britain, the largest of a group of active volcanic islands in the Birsmarck Archipelago, located off the northern coast of Papua New

Guinea and forming part of it politically. It is a mountainous and thickly forested island with an area of 14,600 square miles (23,500 square km); its flightless rail is now uncommon in its main habitat—lowland forest and on the lower slopes of the mountainsides to about 1,500 feet (450 m)—and is even less frequently seen in secondary forest. Its forest habitat is now threatened by logging, and it is trapped and hunted with dogs.

New Guinea Flightless Rail (*Megacrex inepta*)

This rail was first described scientifically in 1879. It is a large, powerful bird with sturdy legs and has been likened to a miniature cassowary in its build and temperament. In defense it kicks, stabs with its beak, and even deters attacking dogs with its aggression. It is totally flightless, but, like other nonflying rails, climbs trees to roost above ground and to escape terrestrial predators. When walking and running it has the distinctive habit of flicking its wings upwards. It is still locally a common bird, in mangrove forest, bamboo thickets, and sago swamps, and although habitat loss and hunting are apparently affecting its already small numbers, the major threat to its continued survival is from feral pigs, which eat its eggs. There are two races of this rail: *Megacrex i. ineptus* lives in the southern half of New Guinea and has a gray forehead, dull red-brown crown and neck and white underparts. *Megacrex i. pallidus* occurs north of the island's central mountain ranges, and differs in that its underparts are mostly pale buff.

Inaccessible Island Rail (*Atlantisea rogersi*)

Inaccessible Island is one of the main islands in the Tristan da Cunha group in the South Atlantic Ocean, and is one of those very rare places in the world which is uninhabited and has no introduced predators. The island is the remains of an eroded volcano, with a land area of 12 square miles (20 square km), which reaches a maximum altitude of 180 feet (55 m). It was discovered by the Dutch in 1652, and so named because the landing party could not penetrate inland past the steep coastal cliffs and densely wooded gullies and ravines. There were two failed attempts at settlement in the nineteenth century, first by an English family and then by German colonists, and it has remained uninhabited since, visited only by scientific expeditions and Tristan islanders. It was recently declared a nature reserve by the Tristan Island Council to protect its millions of seabirds, the rail, and other endemic land animals. However, the islanders can continue to collect driftwood and guano there as they have done traditionally, despite the risk of transporting black rats from Tristan. Fortunately, none of the alien animals that were introduced over the years—sheep, cattle, pigs, and goats—survived to cause the destruction they have on other islands.

The Inaccessible rail is the world's smallest flightless bird, only 5 inches (12 cm) long and weighing just over 1 ounce (28 g), and is rusty-brown in color with a dark-gray belly, black bill, and red eyes. Its ancestors are believed to have been blown from South America because a parasitic fly found on the rails also occurs on birds in Argentina. Its wings are short and soft, useless for flight, and its feathers have discontinuous barbs and therefore have a loose, hairy appearance. It is restricted to the island, and has never occurred on the others in the group.

The rail is most plentiful on the central plateau, especially in the ferns and tussock grass (*Spartina arundinacea*), where it makes runways through the tall clumps that are flattened by the wind to create a natural roof, and also nests under its protection. The grass is so dense the rails often cannot see each other, and call to maintain contact. Much of the island's vegetation, and presumably many of its rails, were destroyed by fire in 1909, but both have since recovered and the rail's current population is estimated at about 8,000 birds. Birds restricted to a single small island are obviously at greater risk than those with a wider range, and the Inaccessible Island rail is therefore considered vulnerable, for although the island is free of carnivores, both native and alien, the risk of its extermination through the accidental arrival of rats or cats is very great. Unlike other rails, it is a slow breeder and lays only two eggs in a clutch, and its tiny chicks are vulnerable to another endemic species—the Tristan thrush (*Nesocichla eremita*).

Woodford's Rail (*Nesoclopeus woodfordi*)

This is a large rail, 12 inches (30 cm) long, with dark-brown upper plumage and sooty-gray underparts mottled with white on the lower abdomen, and the underwings are barred with white. It has short, thick, pale-blue legs and a long horn-colored bill. It originally occurred on all of the major Solomon Islands—Bougainville, Santa Isabel, Guadalcanal, Choiseul, New Georgia, Malaita, and Kolombangara—in lowland forest and abandoned gardens, but suffered from hunting and predation by domestic cats and dogs. Its status varies considerably on each island. The race *N. w. immaculatus* is apparently still plentiful on Santa Isabel; the race that occupies Choiseul, Kolombangara, New Georgia, and Malaita still barely survives, whereas the rail on Guadalcanal, *N. w. woodfordi*, is believed extinct. Before the civil war it was also presumed extinct on Bougainville, the large northern island which is now politically part of Papua New Guinea, as the subspecies there, *N. w. tertius*, had not been seen for years, the last specimen having been collected in 1936. However, in 1999, Don Hadden, a New Zealand ornithologist, visited the capital Arawa one year after the ceasefire that ended the ten-year war between the Bougainville rebels and the Papua New Guinea Defence Force, and discovered the rail living quite openly in the town. The conflict had allowed native grasses, including the 10-foot-tall (3 m) elephant grass to grow in the untended fields and plantations, providing the perfect habitat for rails. The few that had survived had repopulated the area, another example of how the secretive rails can be overlooked under some circumstances. The flying ability of this bird is still somewhat uncertain, but based on observations made on Bougainville in 1985, it is believed to be flightless.

Samoan Moorhen (*Gallinula pacifica*)

Little is known of this small, nocturnal rail, which is just 6 inches (15 cm) long and is believed to be flightless. A somber bird, it has dark-olive upperparts with a black rump, tail, and tail coverts; and slate-blue throat, neck, and breast. This dullness is relieved by its bright yellow frontal shield and red bill and legs. The Samoan Islands in the South Pacific Ocean have been divided politically for over a century, into American Samoa and the larger, self-ruled group of Western Samoa. The Samoan moorhen has only ever been seen on Savai'i—the largest of the Western Samoan islands—which rises to an altitude of 6,000 feet (1,845 m) on Mt. SiliSili and which is still mostly covered by tropical rain forest. It was seen by members of the Challenger Expedition in 1873, who collected two specimens, and was then believed extinct until recently, when it was seen in 1987 in the mountain forest west of Mt. Eleitoga, and again near Mt. SiliSili in 2003. Although dense forest is not typical rail habitat, they may have been forced there by persecution from humans, cats, and dogs on the more level and open coastal regions, where wild cattle and pigs have also destroyed much of the ground cover, which the birds prefer. It has been suggested that it may live in burrows, as rails have been known to enter burrows to escape a threat. Its large eyes also imply nocturnal activity, which would explain why it has rarely been observed in forest that is unlikely to be entered in darkness. Commercial clear-cutting of the forest near where they were last seen is now contemplated.

Snoring Rail (*Aramidopsis plateni*)

Also known as Platen's Celebes rail, this species is a native of north, central, and southeast Sulawesi (formerly Celebes) in Indonesia. It is known from eleven specimens collected in dense secondary growth, bamboo and rattan thickets bordering lowland and montane forest, and from several sightings in Lore Lindu National Park and adjacent areas, where crabs and lizards are apparently its favorite foods. Like most rails it is not a brightly colored bird, with olivaceous upperparts and a gray mantle, white chin and upper throat, while its belly is dusky-brown barred with white and buff. Its name stems from its habit of making a low growling or snoring sound. It is now considered endangered and is believed to be continually declining due to the loss of habitat, hunting, and predation by feral cats and dogs. Although extensive deforestation has in places increased its preferred habitat of dense secondary growth and tall grass, it has been severely affected by introduced predators and hunters.

The snoring rail's survival in Sulawesi's Lore Lindu National Park, which covers a vast area of the island's center, is now considered threatened due to the increasing human encroachment into the park, plus the feral cats that have become established there. Although its numbers are believed to be low, the snoring rail is also a very secretive bird which has been collected scientifically on only a few occasions, and a museum collector in the first half of the last century spent two years acquiring a single specimen. The bird's best hope of survival appears to be on the neighboring island of Buton, where a large tract of protected watershed forest remains, and which is now being recommended for national park status.

San Cristobal Moorhen (*Gallinula silvestris*)

San Cristobal Island, or Makira as it is now called, is the southern-most large island in the Solomon Islands Archipelago in the South Pacific Ocean. Its rail is known scientifically from only one bird collected in 1929 in the island's central ranges by the Whitney South Seas Expedition. The museum specimen shows a very dark-brown bird, with a glossy bluish head and neck, and a greenish-blue forehead shield, red bill, and legs and feet and underparts of slate-gray. The hunter who caught it said that he often saw these birds, in both the primeval mountain forest and in secondary growth and native gardens, but they were more plentiful in the mountains. He said they were hunted with dogs, and did not fly but scrambled into bushes to escape the dogs. Efforts since then to acquire further specimens have been unsuccessful, although the area where it was discovered has not been visited by ornithologists for almost half a century, yet its presence has continued to be reported by local hunters. The volcanic island is still heavily forested on its upper levels, although logging and clearance for cultivation have occurred in the lowlands.

Invisible Rail (*Habroptila wallaci*)

This rail is synonymous with Wallace's rail (*Rallus wallaci*), and is endemic to the island of Halmahera in Indonesia's Maluku province (formerly Molucca). It is a large bird, 16 inches (40 cm) long, with dull slate-gray feathering above and chocolate lower back and tail coverts and shortened black primary feathers, and like most rails its belly is paler, barred with brown. This dull feathering is brightened by its frontal shield, eye-ring, and feet and legs, all of which are bright red. Its voice has been described as a low drum beat interspersed with loud screams. The invisible rail has always been considered a rare species, but has seldom been seen and is known scientifically only from a few museum specimens. It apparently prefers to live in dense alang-alang grass thickets in the sago swamps, especially along the edges bordering the forest, and the local people say their dogs occasionally catch them, but contradictory early reports that it was common may have resulted from confusion with another species, the common bushhen. There had been no confirmed sightings of the bird for forty years, and it was consequently believed extinct, until it was recently observed by a group of ornithologists. The main threats to its continued existence are the destruction of the island's sago swamps and predation from cats and dogs.

Guam Rail (*Rallus owstoni*)

The wildlife of Guam, an island of about 175 square miles (280 square km) in the Marianas of Micronesia, has suffered severely from the introduction of predators, and its flightless native rail was almost exterminated. The island's bird population began to plummet in the early 1960s, with the forest species in the south disappearing first. The decline spread northwards and by 1983 rails occupied only about 370 acres (153 ha) of forest at the north end of the island. In 1984, after three other endemic species were exterminated and only twenty-one rails survived, fifteen individuals were rescued for captive breeding programs. It was soon obvious that the usual reasons for the extinctions of island birds—rats, cats, dogs, the introduction of bird diseases, and the loss of habitat—did not apply to Guam, and the plight of the rail and other birds was blamed on the alien brown tree snake, a native of Australia, New Guinea, and the Solomon Islands. The snake's presence on Guam was first reported in the south in the early 1950s after probably arriving as a stowaway aboard a military cargo plane. As the snake expanded its range, by 1 mile (1.6 km) annually, there was a noticeable decline in the number of forest birds.

Guam Rail   *This flightless rail was exterminated on the island by the introduced brown tree snake, which also eliminated three other endemic bird species. Captive breeding has produced several hundred rails, which have since been returned to Guam where they are kept initially in a snake-proof enclosure, and also to the snake-free neighboring island of Rota.*
Photo: Courtesy National Zoological Park

Guam was the paradise for the snakes that it had earlier been for the rails, because they had neither predators nor competition. As they were basically lizard-eaters the abundance of lizards on the island supported a larger population of snakes than the birds alone could maintain, and therefore provided food guarantees for the snakes as they decimated the birds. The Guam rail became extinct in the wild, but has bred prolifically in several zoological institutions, and reintroduction programs have been successful. The first releases were on the neighboring snake-free island of Rota in the Northern Marianas in 1989, and several hundred have since been introduced. They have also been returned to Guam into a 60-acre (24 ha) fenced area cleared of snakes. Sixteen rails were released there in 1998, and within two years had produced forty-six young.

Auckland Island Rail (*Rallus pectoralis muelleri*)

Restricted to the Auckland Islands, which lie in the Pacific Ocean 250 miles (400 km) south of New Zealand, this rail is a race of the slate-breasted rail (*Rallus pectoralis*), a widespread species in Indonesia and Australia, which was blown to the islands on the west winds. The Maori were the first human visitors to the islands from the neighboring Chatham Islands, and although discovered by westerners in 1806, they were not colonized by the English until 1842, and from then on they were plagued with introduced animals. Pigs were released as food for visiting mariners and castaways; and cattle, rabbits, sheep, rats, cats, and dogs came ashore with the settlers. Combined, they had a disastrous effect on the distinctive endemic fauna. The rail, like the island's flightless brown teal, was soon exterminated on Auckland Island and now survives only on neighboring Adams Island, which has a land area of 38 square miles (61 square km), and on tiny Disappointment Island, which is just 1½ square miles (2.4 square km) in extent. Its wings are greatly reduced, but while it is believed to be flightless, this has not been reliably confirmed.

The attempt at settlement failed and the islands were vacated in 1852, leaving the cattle and sheep behind, but they soon disappeared too and are believed to have been eaten by the Maoris. Further efforts to colonize the islands were also unsuccessful and they are now an uninhabited nature reserve. The uncharted waters around the islands were a serious threat to vessels, many of which foundered on their rocks. There are no alien animals on the islands now, and apart from some changes to their original vegetation from the pigs' rooting years ago, they remain largely unchanged. Scrub and forest, mostly of rata (*Metrosideros umbellatus*), fringe the islands, and tussock grass (*Chionochloa antarctica*) grows on the upper, level areas. The rail currently numbers about 400 on Disappointment Island and 1,500 on Adams Island.

Gough Island Moorhen (*Gallinula nesiotis comeri*)

A World Heritage site, with many endemic plants and thousands of seabirds, Gough Island is considered one of the least-disturbed cool-temperate islands in the world. It is one of the Tristan da Cunha group, lying in the middle of the South Atlantic Ocean, 230 miles (370 km) southeast of Tristan. Discovered accidentally by a Portuguese navigator blown off course while attempting to round the Cape of Good Hope, it was later named after Captain Gough who rediscovered it in 1732. A small island of just 25 square miles (40 square km), it has an isothermal or almost constant mild temperature, but it is steep and rugged, rising to 3,000 feet (915 m) at Edinburgh Peak, and is protected by steep sea cliffs up to 1,500 feet (458 m) high, with few safe anchorages, so it did not encourage human settlement. It is now a wildlife preserve, on which unapproved landing is prohibited, and is still uninhabited except for the staff of a South African–manned weather station which has been there since 1956.

Gough Island lies in the path of the roaring forties, the gale-force winds that blow in a clockwise direction around Antarctica, so it is exposed to frequent storms and heavy rainfall. In such an environment flightlessness was an advantage as it prevented birds from being blown out to sea during violent storms. George Comer, second mate of an American sealer that visited the island in 1888, wrote the first account of the endemic moorhen—a rail despite its name—which lives in the tussock grasslands on top of the island. It reached the island group long ago and was believed to have diverged into two subspecies: one on Gough Island and the other on neighboring Tristan da Cunha which was exterminated in 1890; but it is now thought that they were the same and not distinct races. The Gough Island moorhen is definitely flightless and can just flutter a few feet on its short wings. It has no known predators, but the estimated 3,000 breeding pairs of Tristan skuas on the island may well be a threat, as they prey heavily on the eggs and young of the rockhopper penguins nesting there. The moorhen's current population is about 5,000 on Gough Island, and perhaps 500 on Tristan da Cunha, to where it was reintroduced in the 1950s, and where, as an alien species, it is not protected by the island's conservation laws. In turn it has been accused of eating the eggs of the yellow-nosed albatross (*Diomedea chlororhynchos*), of which most of the world's population nest on Gough Island.

Lord Howe Island Rail (*Tricholimnas sylvestris*)

A cresent-shaped island, just 6 miles (9.5 km) long and little more than 1.2 miles (2 km) wide, the World Heritage Site of Lord Howe Island lies in the Pacific Ocean 500 miles (800 km) northeast of Sydney. Administratively it is part of New South Wales and is now a popular tourist destination. It is a distinct part of the Australian ecozone, with a large endemic flora and fauna, including the kentia palm (*Howea forsteriana*), a familiar house and conservatory plant in the Northern Hemisphere. As Lord Howe Island was never attached to mainland Australia, its original life forms have arrived naturally by air or sea. Unfortunately, since its discovery by the Royal Navy in 1788, and its permanent settlement in 1833, a host of other forms of animal life have also arrived, to the detriment of the native species. Several of the island's birds are extinct, including the flightless Lord Howe Island swamphen or white gallinule, the Lord Howe Island gerygone (a warbler), and the tiny robust white-eye, all falling prey to the introduced rats.

The island's vegetation has also been dramatically altered. Much of its original forest has been destroyed, cleared for agriculture or seriously disturbed by alien animals. Feral goats ate leaves and bark and climbed into the tree canopy to feed; and in destroying seedlings they also affected the forest's chances of rejuvenation. Pigs rooted through the soil, eating plants and disturbing new growth, and were also partial to the eggs and young of ground-nesting birds. Feral cats were also a problem, as were rats descended from those shipwrecked in 1918, and the small native flightless rail was soon in trouble. In 1963 its population was estimated to be about 200 pairs, but by 1980 it had been reduced to about 20 birds. A breeding program was initiated by the Australian National Parks and Wildlife Service, and three pairs were caught and established in captivity. They responded spectacularly and raised thirteen chicks in their first season. Two more pairs were captured the following year, and by its end seventy-eight chicks had been raised. Most of these were returned to the island and by 1984 the population there had grown to over 200 birds, and bird watchers now see rails regularly in several locations on the island. Goats have since been eliminated, the feral pig population has been reduced, and rats are being controlled.

Not all bird rescue and captive breeding efforts have been as succeessful as the Lord Howe Island Rail project, but its success illustrates the potential value of captive breeding, endorsed by both the World Conservation Union and the World Wildlife Fund as an integral aspect of conservation when it is obvious that the rapid decline of a species foretells its imminent demise. Together with the more recent success in breeding the Guam rail, it shows how receptive rails can be to a properly conceived and managed breeding program.

New Caledonia Wood Rail (*Gallirallus lafresnayanus*)

The flightless rail of New Caledonia, a large island in the South Pacific west of Fiji, is a very similar bird to the Lord Howe Island rail—a large, brown rail with dark-gray head and grayish-brown underparts. It is believed to be nocturnal and is a very fast runner, and when alarmed it squeezes into holes and crevices and feigns death. Consequently, it could not cope with the introduced predators—cats, rats, dogs, and pigs—and until recently was believed extinct as it had not been seen since 1890 and had already appeared on lists of vanished birds. But as we have seen on a number of occasions, it is premature to write off forest-dwelling rails based on the lack of sightings, as they are normally very secretive and when threatened have adopted nocturnal habits. New Caledonia is frequently visited by bird-watching groups, especially hoping to see the flightless kagu, and optimism that the wood rail may still survive was recently rewarded with several possible sightings in the north. Although unconfirmed, they have given hope that the species may have survived in the isolated pockets of humid forest that still remain, for the island has been extensively logged.

Zapata Rail (*Cyanolimnas cerverai*)

The Zapata rail is about 11 inches (18 cm) long, with dark olive-brown and slate-gray plumage, and with a conspicuous white throat and white undertail coverts. Like many rails its feet and legs are its most colorful features—being bright red—and its bill is red at the base and yellowish-green terminally. It has short wings and its flight is very weak and limited, but it may not be totally flightless.

The Zapata rail is now an extremely rare bird in serious need of immediate conservation, known to occur only in two areas of Cuba's Zapata Swamp, mostly in the vicinity of Santo Tomas. It is seldom seen, but in 1994, Spanish and Cuban ornitholigsts saw several in two separate areas, one of them a new location for the species. It was concluded that the bird probably migrates locally according to the seasons—wet or dry. Its breeding biology is unknown as its nest has never been found. Burning the dense, bush-covered swamps during the dry season has reduced its available habitat, and introduced predators have aided its decline.

South Island Takahe (*Notornis mantelli hochstetteri*)

The takahe is the most impressive member of the rail family, a gallinule that lost its ability to fly long ago and grew very large, due to its formerly secure habitat and the lack of predation and competition. It is believed to have evolved from an early invasion of the purple swamphen from Australia, probably not long after New Zealand broke away from Gondwanaland. Later arrivals of the same species gave rise to the pukeko, a common bird that is still the same size as its continental ancestor, and still flies. Two races of the takahe evolved, but the North Island one is known only from subfossil bones. The South Island takahe is the size of a chicken, males weighing almost 6 pounds (2.7 kg) and females slightly less, but still resembles its ancestors in color, with deep purplish-blue head, neck, and underparts, and bronzy-green back. In keeping with its size it has very thick legs and a heavy bill and is now totally vegetarian, surviving mainly on the succulent shoots of three species of snow tussock grass.

The first recorded live takahe seen by Europeans was caught by sealers' dogs on South Island in 1849. It was kept alive for several days, then roasted and eaten, being declared delicious. Others were caught in 1851, 1879, and 1898, but half a century then passed without sightings or the discovery of a single takahe, and it was considered extinct. However, in 1948, Dr. E. B. Orbell discovered a small population in an isolated valley in Fiordland's Murchison Mountains, west of Lake Te Anau. Studies by the Department of Conservation have since shown that the takahe occurs over an area of almost 115 square miles (185 square km) of mountain slopes and valleys, where the vegetation is mostly tussock grass and scrub. Its reduction to this current small range is believed to be due to hunting by the Maoris and their dogs, overgrazing by introduced red deer, climate change, and possibly disease.

The takahe is a sedentary bird, with pairs remaining in the same territory all year, and not migrating to lower elevations for the winter. It nests in the tussock grass, laying one or two creamy-white eggs with brown and purple blotches, which are incubated for thirty days. When rediscovered, its population was estimated at about 250 pairs, but by 1970 only half this number survived. As a result of the captive breeding program at the National Wildlife Center, and especially the removal of an egg from all nests in the wild that contained two fertile eggs (for artificial incubation and hand-raising), birds have been released into other suitable habitat. The takahe is now also established in Fiordland's Stewart Mountains, and on four

Takahe   The largest rail, a chicken-sized descendant of Australia's purple swamphen or gallinule, possibly isolated in New Zealand when it separated from Australia soon after the breakup of the super continent Gondwanaland. It is the only totally vegetarian rail, and survives on the succulent stems of snow tussock grass.
Photo: Clive Roots

islands—Kapiti, Mana, Maud, and Tiritiri Matangi. Its current population is almost 200 birds, with about 150 in the wild in Fiordland, 34 on the predator-free islands, and 12 in captivity. Its ability to thrive in habitat other than the tussock grassland, where it was confined at the time of its rediscovery, is proof of its original wider range, its survival only in the remote and uninhabited valleys of Fiordland being due to the lack of the usual pressures associated with colonization. (See the color insert.)

Calayan Rail (*Galirallus calayanensis*)

The Calayan rail was discovered by chance in August 2004 on the remote island of Calayan in the Babuyan Islands off the northern tip of the Philippines, by a team of Filipino and British zoologists. The bird was well known to the inhabitants, but the island had not been visited by ornithologists for a century. It is a dark-brown bird, the size of a small crow, with an orange-red bill, legs, and feet, and with short wings. It was observed from sea level up to 1,000 feet (300 m) on the island's forested slopes, usually in the vicinity of streams. The rails were never seen to fly and were therefore thought to be flightless or nearly so, and the bird that was killed and dissected for scientific recording purposes was found to have very weak flight muscles. Its population was estimated at between 100 and 200 pairs. Although the new bird is not under any immediate threat, the expedition team plans to undertake further research to determine the requirements for its long-term preservation.

Note

1. A rather misleading name that dates from the time when mor meant a morass—a bog or marsh—not a heather-clad moorland.

4 The Gliders

With just two exceptions all the flightless land birds are either ratites or rails. The exceptions are the kakapo and the kagu, two of the world's rarest and most unusual birds, which have both reached the stage where they still have large wings but cannot become airborne. The degeneration of their breast muscles and flattening of the keeled breastbone, which are essential for flapping flight, have restricted them to gliding only, the kagu along the ground after a short run, and the kakapo back to the ground from a tree it has climbed.

The Species

Kakapo (*Strigops habroptilus*)

The kakapo is the world's largest and most unusual parrot, weighing 8 pounds (3.6 kg) and measuring 25 inches (63 cm) long. It has a facial disc of bristle-like feathers and its soft, owl-like plumage is bright green with yellowish underparts, all barred and streaked with brown and yellow, and with a yellow stripe over its eye. (See the color insert.) It is a nocturnal woodland bird, originally occurring from sea level to 4,000 feet (1,220 m) and preferring mossy beech forests, especially where they adjoined snow tussock meadows or river flats. During daylight it shelters in burrows, crevices, among tree roots or in dense low vegetation, and is only occasionally seen abroad. A fast but awkward runner, after dark the kakapo follows well-defined trails to its feeding grounds in the subalpine zones, often up to 6 miles (9.5 km) from its daytime roost.

Kakapos are almost totally vegetarian, like all the members of the parrot family *Psittacidae*.[1] They eat a wide range of native and introduced vegetation including shoots, leaves, nuts, berries, fern roots, fungi, moss, and occasionally they find

Kakapo *New Zealand's very rare parrot the kakapo, the world's largest psittacine, cannot fly despite its large wings (although with shortened and rounded primaries), as its flight muscles have completely degenerated and its breast keel is now just a low ridge.*
Photo: Courtesy Department of Conservation, New Zealand. Crown Copyright. Photographer: Don Merton, 1999

insect larvae. They browse by pulling a leaf through their bill with a foot, drawing out the nutrients and leaving balls of masticated fibrous cellulose hanging from the living plant, and making an easy trail to follow. Despite its large, strong wings, which appear adequate for flight, the kakapo can no longer fly because its flight muscles have completely degenerated, and its keel is now just a low ridge. It climbs well, however, after the fashion of the parrots, using its feet and bill and balancing with outstretched wings. When returning to the ground from a high point it spreads its wings and glides down, but the wings barely slow its descent and it often hits the ground with a thump. In captivity it has not been observed to climb and uses sloping branches to gain height.

Kakapos make deep booming calls and discordant shrieks and croaks. The shallow, cleared depressions that have been found near their booming grounds may be associated with their display, which is believed to be a communal lek—a form of polygamy where several males congregate to display to the hens that visit them only for mating. They nest in midsummer, in a burrow beneath tree roots or in a rock crevice, lining the nest with decaying wood and feathers. The slow rate of breeding has not improved their survival chances, as they do not breed annually, and do not reach breeding age until they are at least eight years old. Two eggs are usually laid and, in keeping with the assumed lek display of the male, it is believed that only the female incubates the eggs and raises the chicks. From observation of nests that contained a down-covered chick and an almost fledged youngster, it is also assumed that there is a long interval between egg laying, and incubation begins when the first egg is laid.

Kakapos were once very common birds in the forests of North, South, and Stewart Islands, and although their downfall began with the arrival of the Polynesians, they were still plentiful in the early days of European settlement. While exploring the west coast of Otago in 1861, Sir James Hector "found the kakapo very numerous," and the Maoris told magistrate Sir Walter Buller in 1870 that kakapos gathered in caves during the winter in such numbers that their shrieks and harsh croaks were deafening. He said they had a thick layer of blubber on the breast, below which the flesh was white and good eating, despite its strong flavor. The Maoris considered them a delicacy, and hunted them at night with dogs and torches. Unfortunately, the dogs so enjoyed kakapo hunting that they also did so "largely on their own account," becoming very fat and soon exterminating all the birds in the vicinity of a camp. Maori hunters and the packs of feral dogs that roamed New Zealand in the early nineteenth century undoubtedly had a drastic effect on the kakapos.

European settlement placed additional pressures on the kakapos. Land clearance, logging, introduced predators and herbivores, and bird diseases, combined with the Maori hunters and their dogs, and rats, stoats, and feral cats were especially destructive. Introduced deer also assisted the parrot's demise, through overbrowsing the vegetation and trampling the trails that led up to their feeding grounds, and kakapos only survived in valleys where deer had not penetrated. Based mainly upon their size, three races have been described, *Strigops h. habroptilus* from South island, *Strigops habroptilus parsoni* from the southern alpine regions, and *Strigops habroptilus innominatus* from North Island, but this "splitting" of the species is not generally accepted. The last sighting of a kakapo on North Island was of a single bird in the Huiara Range in 1927. It was apparently still plentiful on South Island's west coast at the turn of the century, but observations ceased there in the mid-1920s, and after one was caught on Stewart Island in 1949, the kakapo was believed extinct. However, like the takahe, rediscovered years after being presumed lost, the kakapo also survived. A small population was discovered in 1958 in the Cleddau watershed of Milford Sound in Fiordland. In 1976 more were found on Stewart Island and in a remote region of Doubtful Sound, and in 1985 even in northwest Nelson, far removed from the southern populations.

The first major attempt to preserve the kakapo was made at the end of the nineteenth century, when Richard Henry, curator of Resolution Island, New Zealand's first gazetted bird sanctuary, caught 400 birds in Fiordland and transferred them to islands in Dusky Sound. Stoats, introduced into South Island from Europe, swam across from the mainland and killed them all, a most unfortunate experience but one that influenced New Zealand's many subsequent animal translocations. When the Kakapo Recovery Plan was formulated in 1989 there were just forty-three known surviving birds. Relocating them to safe islands has been a major aspect of their conservation program for some years, their sex being determined by fecal hormone analysis before they were moved. Further study has shown that the males have broader heads and bills than the females, and their eyes are wider spaced.

In view of the great danger posed by cats on Stewart Island, twenty-two ka-kapos were moved from there to Little Barrier Island in 1982. Three years later only two chicks hatched from the nine eggs laid by three females on Stewart Island, and they were killed by cats. Consequently, between 1987 and 1992 all the known kakapos on Stewart Island, thirty-seven of them, were caught and moved to pred-ator-free Codfish (now called Whenua Hou), Mana, and Maud Islands, although the latter translocation failed. To improve the kakapo's chances of survival the alien brush-tailed possums were eradicated on Whenua Hou in 1984, and rats in 1999. Chicks also vanished from their nests on Little Barrier Island in 1993, apparently due to depredations by Polynesian rats. Once thought to be relatively harmless, at least compared to the alien black and brown rats, it is now evident that the Polynesian rat eats the eggs and young of ground-nesting birds, and attempts are being made to eradicate the rats on Little Barrier Island. The kakapo was extinct on South Island by 1989 and with the relocation of all the Stewart Island birds it no longer occurs in its natural range. In New Zealand it was the first time that an entire species had been relocated.

One of the major problems facing the kakapo's recovery is the disproportionate ratio of hens to cocks, with only thirteen hens in the total population of fifty-four adult birds. Until recently the only natural breeding to occur since 1981 was the raising of a single chick in 1992, but after an intensive program by the Department of Conservation, eleven eggs were laid and three chicks hatched early in 1997 on Whenua Hou Island. There followed a period of several years with little increase in the parrot's population, until February 2002 when the Department of Conservation reported that fifty-two eggs were being incubated by eighteen females on the island, and the first chick hatched on February twentieth. In total that year the kakapo population was boosted by twenty-two chicks, increasing their number to eighty-four. The excellent breeding results are attributed to the fruiting of the rimu or red pine (*Dacridium cupressinum*), which is plentiful on the island, as its fruit forms almost the sole diet of both adults and their chicks and greatly improves their breeding success rate. The rimu is a dioecious tree—one in which the male and female reproductive organs are in different trees—which grows at low elevations in mixed forests throughout New Zealand, generally reaching a height of about 100 feet (30 m), although specimens of almost 150 feet (45 m) are known. Captain Cook fermented its young shoots with sugar as a remedy for scurvy on his second voyage in 1773. Unfortunately, the rimu seeds only at irregular intervals, generally every five years, and is such an important source of food for the kakapo nestlings that the parrot's most successful breeding years coincide with its fruiting.

In the past decade the Kakapo Recovery Program has used some well-tried conservation techniques and procedures, and has developed many new ones. All the birds are fitted with radio transmitters, and infrared cameras monitor the nests during the breeding season. Nesting birds are watched continually, and the eggs or chicks are covered with a warming pad when the hen leaves them to feed. Eggs have also been fostered for incubation to females that did not have their own eggs; and chicks that do not look well are removed for hand-raising. The wild bird's natural diet is supplemented with a special mixture of cereals. But even now, with all the accumulated knowledge of the bird and its maintanenace and breeding requirements, there are still disappointments. Thirty-one kakapos were transferred by helicopter in July 2004 from Whenua Hou to predator-free Te Kakahu Island in Fiordland, but septicaemia resulting from an infection of the soil bacterium erysipelas claimed the lives of three birds soon after their arrival. The others were immediately treated with antibiotics and were then vaccinated, as were those remaining on Whenua Hou.

In August 2004 the kakapo's population numbered eighty-three. All birds are known individually and their life histories are well recorded. The oldest one, named Richard Henry after the pioneer conservationist who attempted to save them over a century ago, was caught in Fiordland in 1975 and is now estimated as being fifty years old, but he has sired chicks recently. He is so tame, he has fallen asleep in people's arms.

Kagu (*Rhynochetos jubatus*)

The kagu is the national symbol of New Caledonia and is restricted to that island. It is a unique and distinctive bird, the size of a small chicken, about 22 inches (55 cm) long, and the sexes are alike externally, with ashy-gray plumage and a crest of long feathers hanging down the nape, which is raised in display. Its wings are barred with white, brown, and black, but this patterning is only visible when they

Kagu *The chicken-sized kagu is related to the cranes and rails and is endemic to the southern Pacific Ocean island of New Caledonia, of which it is the national symbol. Completely flightless due to the deterioration of its breast muscles, it still has large, broad, and rounded wings on which it glides for long distances down the mountain slopes. It is one of the world's rarest birds with a population of less than 1,000.*
Photo: Courtesy Kenneth W. Fink

are opened. The kagu has strong, reddish-orange legs and a powerful bill. Locals call it the "ghost of the forest" and it is a very rare bird that is included in Appendix II of the International Convention on Trade in Endangered Species.

New Caledonia is a long, narrow island northeast of Australia in Oceania, with a land area of 45,000 acres (18,575 ha), which rises to the peak of Mont Panie at 5,350 feet (1,628 m). It has a humid, tropical climate and was formerly heavily forested. Settled by both the British and French early in the nineteenth century, it was made a French possession in 1853, served as a penal colony until the end of the century, and is now an overseas territory of France. Violence connected with the desire for self-government erupted in the 1980s and early 1990s, and it was during this period that the fate of the kagu did not look promising, particularly as so many Europeans were leaving the country.

A member of the order *Gruiformes*, which contains the rails and cranes, the kagu is a monotypic species—the only one of its kind—and in a family of its own, the *Rhynochetidae*. Its closest relative is the sun bittern of South America, and its breast muscles have deteriorated to the extent that its relatively large, broad, and rounded wings are useless for flight. It lives in forest and scrub, and when threatened it moves fast on its chicken-like legs, running down the mountain slopes and launching into a long and low graceful glide away from danger. It is certainly the best glider of all the flightless birds.

The kagu's diet consists mainly of earthworms, snails, and small lizards. When searching for worms it taps the ground like the kiwi to locate them, and then digs them out with its strong beak. It was originally believed to be nocturnal as its loud, piercing calls were heard in the forest at night, but recent observation has now proved that it is active during daylight. The kagu has a harsh rattling call and a more melodius predawn song, and is semicolonial in its habits, forming loose flocks throughout the year and then pairing off at breeding time. The small flocks may be composed of a breeding pair and their young, which stay with the parents for several years, helping to raise the chick of the year. The kagu's nest is a simple layer of leaves on the forest floor, and a single creamy-yellow egg with reddish and gray blotches is laid in the dry season between May and December. The egg is incubated for 33–35 days, and the newly hatched chick is a pretty little bird, covered with black-and-yellow streaks, which can run rapidly when 15 days old. Kagus have always been rare in zoological gardens, but the few specimens they have acquired lived very long lives. An individual lived for twenty-one years at the West Berlin Zoo, where the first captive breeding occurred in 1964, exactly one century since the first kagu was imported into Europe. They have been bred more recently at the San Diego Zoo.

The kagu evolved on an island free from predation, and its first encounter with a predator intent upon killing it occurred when the early Melanesians settled on the island and trapped it for food. They were followed many years later by the European colonists who also hunted it for food and then early in the last century caught many birds for the plume trade, using dogs to locate and flush the birds out. The kagu population also suffered from all the usual threats to island birds such as habitat disturbance and loss, in New Caledonia's case the felling of the *Araucaria* pines; predation by cats, rats, and pigs; and the packs of feral dogs that roamed the island.

At the International Union for the Conservation of Nature's General Assembly at Madrid in 1984 the kagu was one of twelve animals identified as the world's most endangered species. Two decades later it is still endangered but is unlikely to qualify for such a distinction now, not because of greatly increased numbers, but because there are now many more species with smaller populations and at greater risk.

The kagu's future appeared bleak when the political violence ended, as protection for the bird was minimal and plans to create a national park had failed, but Yves Letocart took over the bird's conservation in 1980 and the kagu's fortunes changed. A New Caledonian of French descent, he began a predator-removal campaign in the area known as Riviere Bleue, which then contained perhaps sixty birds. In 1989 he began radio-collaring kagus to improve the knowledge of their movements, territory, and breeding biology. Among the extensive new information he discovered on their breeding biology was the fact that young birds stayed with their parents and assisted in raising future chicks; young females then left to establish their own territories at the age of two or three years, while the males stayed with their parents for a few more years. It was Letocart who discovered they are not nocturnal, but just awaken to call in the night. His successful predator-control program has been directly responsible for the increase in birds at Riviere Bleue, now a National Park of which Letocart is the superintendant, to approximately 300. Fortunately, dogs have been eliminated from the park, for elsewhere they are the kagu's most serious threat, having killed seventeen radio-collared birds in the Pic Nigua Reserve in 1993.

The kagu now receives full government protection and is currently believed to number between 700 and 1,000 birds, surviving in the few remaining tracts of mountain forest on New Caledonia's eastern side. It lives mainly in poorly accessible and therefore unexploited valleys from 1,300 feet (400 m) to 3,300 feet (1,000 m) in the vicinity of Mt. Do and Mt. Nakada, where 135 birds have been counted, and in the Riviere Bleue National Park where there were estimated to be 300 birds in 2002. Many kagu in the park are banded and radio-collared, but the species' recovery from near extinction has been hindered by its slow breeding rate, just a single chick annually per pair at the most. A captive breeding program has been operated by the Southern Province and the Societe Caledonienne d'Ornithologie since 1978 and has produced numerous birds that have been released in Riviere Bleue. (See the color insert.)

Note

1. Except the kea, whose carnivorous tendencies are probably unnatural and were encouraged by the availability of sheep in the past two centuries.

5 Wings Underwater

■ PENGUINS

Penguins are the most recognizable wild birds. They stand upright, have long flippers instead of wings, are torpedo shaped, and walk with a waddle. The eighteen species differ only in their size and facial markings, for they all have white underparts and dark backs that camouflage them from both below and above when they are in the water. Externally, the sexes are also similar, and only the six species of crested penguins and the blue penguin are sexually dimorphic, the males being slightly larger than the females and having more robust bills.

Penguins range in size from the blue penguin, which weighs just over 2 pounds (900 g) to the emperor penguin, which may weigh 65 pounds (30 kg), but most are in the 7–11-pound (3–5 kg range). They have short, thick legs which are set far back on their bodies, four long toes of which three point forward and one backward, and strong claws. Their powerful fish-grasping bills are deep, short, and narrow. They are all marine birds that spend most of their lives at sea, often far from land, and in the water they are the most accomplished swimmers of all birds. Whereas all other flightless species merely lost the use of their wings, the penguin's wings are modified as paddle-like flippers, making them the most specialized of the birds that no longer fly. On land they stand upright and either waddle or bounce along with their feet together, but on smooth ice they may toboggan on their stomachs.

Origins

After years of controversy it is now accepted that the penguins evolved from flying ancestors as their flippers are undoubtedly modified wings; because of the fusion of the wrist and hand bones—the carpals and metacarpals—which is a modification of flying birds for the attachment of the quill or flight feathers; and to withstand the pressures and stresses of flight. The penguin's breastbone is keeled like that of the flying birds for anchoring the muscles that were once necessary for airborne flight and are now needed for underwater "flight." Penguins have a py-gostyle, which is the fused caudal vertebrae and its fleshy covering from which the tail feathers grow. This is an adaptation for flight, with the fan-tailed feather arrangement arising from it, enabling the flying bird to maneuver. Although the penguins have lost these feathers the pygostyle remains as proof of their original flying ability. Penguins sleep with their bills tucked behind their flippers in the typical position adopted by birds with developed, feathered wings, which is also believed to be a legacy of their flying ancestors. Finally, they have a well-developed cerebellum, that part of the brain necessary for the balance required for flight and for coordinating the flight muscles.

The penguin's closest ecological counterparts are the auks of the Northern Hemisphere, which are believed to have evolved from gull-like birds about 85 million years ago. Like them they may have passed through a stage of flying and swimming during their evolution. But the current penguins and the original "pinguin"—the flightless and now extinct great auk—were not related and evolved from diverse ancestors. The great auk, as its name suggests, evolved from the auks, which belong to the same order of birds, the *Charadriiformes*, as the gulls and plovers. Fossils, including one of a penguin almost 5 feet (1.5 m) tall that lived in New Zealand long ago, show that the penguins developed in southern latitudes, similar to their present-day distribution. Although their ancestral origin is still uncertain, they are believed to have evolved from the tubenoses—the albatrosses, fulmars, and petrels—so-called because their nostrils are concealed in a double or single tube on the upper mandible, in a process which began at least 80 million years ago. This is rather surprising when the behavior of the two groups is compared, for the tubenoses are among the greatest fliers in the bird world, soaring for months over the open oceans. However, fossils from the Eocene Period, about 70 million years ago, have features of both the penguins and the tubenoses, and in their modern-day behavior there are still similarities between the two groups. They are both ocean dwellers that spend months on the open seas; some are burrow-nesters and their young are slow to reach maturity. Also, the young of the most primitive species—the blue penguin and the white-flippered penguin—have nasal tube openings similar to those of the tubenoses. There are also similarities in their courtship dances and in their use of stones in courtship, although such comparative behavior may be adaptive, having evolved in response to a similar environment, and is therefore not generally considered valid in bird classification.

Flight was therefore less important to the early penguins than speed and agility in the waters rich in krill, squid, and fish, and over the years they became more adept at swimming and diving, and evolved smaller wings, since this reduced underwater drag. Eventually, they lost the ability to fly and their wings evolved into paddles, powered by the pectoral muscles originally developed for flight and now used for rapid underwater propulsion. Penguins are perfectly adapted for an aquatic life, with most species spending months at sea and only coming ashore to breed and moult. R. C. Murphy, in his encyclopedic *Oceanic Birds of South*

America, sees the penguins as having three successive ancestral stages, "first the flying, second the flightless, third the supremely aquatic."

Range

Penguins live in the Southern Hemisphere, where their distribution is controlled by water temperature. They occur only in the cool waters of the Maritime Antarctic and the southern oceans, from the polar ice cap to the equator on South America's west coast, and into the subtropics on the coasts and offshore islands of South Africa, Australia, and New Zealand. They are the dominant birds of southern waters, and several species range widely. For example, king penguins ringed at Crozet Island in the southern Indian Ocean have been recovered on MacQuarie Island south of New Zealand, over 4,300 miles (6,900 km) away. Rockhopper penguins visit beaches in South Africa, 1,700 miles (2,700 km) from their nearest nesting rookeries on Gough Island in the mid-Atlantic, and vagrant chinstrap penguins have been seen on the Tasmania coast, almost 3,000 miles (4,800 km) from their nearest breeding colonies on Heard Island in the southern Indian Ocean. Excluding the emperor penguin, the Antarctic and sub-Antarctic penguins migrate northward into slightly warmer waters at

King Penguin *A king penguin displays on a South Georgia, Antarctica beach. With a circumpolar distribution this species breeds on many sub-Antarctic islands and currently has a population of about 1 million. On some islands it has never recovered from the massive slaughter of the nineteenth century when these penguins were boiled for their oil.*
Photo: Tony Hathaway, Dreamstime.com

the edge of the Antarctic Convergence for the winter months, and the southern populations of even the temperate climate Magellanic penguin migrate north along the coast of South America for the winter. Only the tropical and warm temperate species are sedentary, for they have no need to escape unfavorable weather.

Although as a group the penguins are the most plentiful flightless birds, the eighteen species vary considerably in their numbers and their survival potential. Whereas some are endangered and at risk of becoming extinct, others are among the world's most numerous bird species. The Fiordland crested and yellow-eyed penguins each have total populations of less than 2,000 birds, only about 1,500 Galapagos penguins remain, and the white-flippered penguin numbers about 4,000. Others have tremedous populations. Adelie, chinstrap, macaroni, and rockhopper penguins number in the millions, and some populations have even increased in recent years due to the reduced competition for food from krill-eating whales; but the increasing interest in krill for human consumption will no doubt eventually reverse that situation. At some nest sites on the Antarctic peninsula the chinstrap penguin has quadrupled its numbers and the adelie penguins have doubled, although elsewhere they have declined. Generally, however, the populations of all species, especially the more plentiful forms, do not compare with the vast flocks of a century ago. The king penguin was virtually eliminated on some islands when it was boiled for its oil. In the twentieth century, the black-footed penguin was reduced to just 10 percent of its original population estimate of 3 million birds, and the Humboldt penguin declined drastically. Both are now included in the Appendices of the Convention on International Trade in Endangered Species of Wild Fauna and Flora (CITES), which monitors or prohibits trade in many wild animals.

Physiology

The penguin's most important adaptations are concerned with insulation and swimming. In most birds the feathers grow in regular groupings or tracts with intervening bare spaces covered by the overlying feathers, but penguins lack these tracts and their very specialized feathers cover the whole body surface. They are short, stiff, and oily, with up to seventy per square inch and their ends turn back toward the body, forming a tight covering which is not ruffled by strong winds. Each feather also has a fluffy aftershaft near its base which forms an insulating layer next to the blubber. Their legs are feathered down to their ankles, and their flippers are covered with short, close-fitting, scale-like feathers. Penguins can control their feathers to lie tight against their bodies, expelling air to facilitate underwater swimming, or they can fluff them up to retain air bubbles, which increases their bouyancy for swimming on the surface. Both methods also help the penguins control their temperature on land; but all species, especially the Antarctic ones which are resistant to very low temperatures, still have difficulty dissipating heat on warm days. They also have a vascular heat-exchange system in their legs and flippers, where arteries carrying blood at body temperature run alongside the veins that are returning cooled blood to the body. This warms the returning blood and holds the temperature of the extremities above freezing. Even penguins living in temperate and tropical latitudes have these modifications, which are useful when the birds are in the cold water, but can result in overheating when they are on land. To compensate, they are able to dissipate more heat than their Antarctic counterparts through their larger flippers and bare facial areas. Many species nest in burrows, in the undergrowth or forest, which also provides protection from the sun.

The flippers are also used to help reduce body temperature. When this is normal a penguin holds its flippers to its sides, but when it begins to overheat it lifts them away from its body. The erect-crested penguin, which nests on bare rocks on the Bounty and Antipodes Islands, east of New Zealand, with no shelter from the sun, has the largest flippers in relation to its body size. A penguin's body temperature varies very little whether it is a Humboldt penguin nesting on a tropical beach in Peru, or an adelie penguin incubating its egg on the Antarctic ice. The former is concerned with losing heat to prevent overheating to fatal levels, and the latter needs to maintain its body temperature in the cold environment. A thick, insulating layer of blubber just below their skins helps the cold-climate penguins reduce their heat loss. In the Antarctic species this blubber is so thick it forms almost one-third of their total body weight, and almost resulted in their demise when they were regarded first as a fuel for rendering seal blubber, then as an alternative to the seals for the supply of oil. Their blubber not only conserves body heat but also acts as a source of energy, especially during the long nonfeeding periods of the nesting season. Conversely, when there is a risk of becoming over-heated, blood vessels in the penguin's skin become engorged, allowing body heat to escape through the blubber and skin to the surface. Antarctic and sub-Antarctic species have feathered legs, but the warm-climate penguins have bare shanks and lose heat through their legs and feet. Their facial patches, which are also un-feathered, can be flushed with blood to help them cool down.

Good underwater vision is obviously important for a bird that chases fish, but although aquatic animals are generally shortsighted on land, penguins also have good vision above the water. The penguin's eye has a flattened cornea, which alters the angle that light can enter the eye; this is important underwater where light enters the eye obliquely.

Practically all the knowledge of penguins comes from observations made on land when they congregate at their nesting rookeries, or in zoos and oceanariums where their swimming ability can be observed on the surface and underwater. They are such superb swimmers that the early mariners navigating the Cape of Good Hope, who were the first Europeans to see penguins, thought they were feathered fish. Even early in the nineteenth century some scientists considered them to be a link between fish and birds. Penguins are the only living birds that power their swimming with wings modified as flippers or paddles, and in historic times the only other bird with these modifications was the now extinct great auk. Other flightless waterbirds such as the Galapagos cormorant and the giant coot use their legs and feet for propulsion. The underwater swimmers that can still fly, such as the auks and diving petrels, also use their wings underwater, with the same short, fast wing-beats for underwater propulsion as when they are airborne.

In penguin "flight" the leading edge of the wing is raised on the upstroke and lowered on the downstroke, providing thrust both up and down and acting as a hydrofoil. Flying through water requires even more power than flight through the air because water is a denser medium, and the penguins have well-developed pec-toral muscles and a keeled sternum. As they are lighter than water they must over-come the upward force, and the penguins' wing bones are broad and flat to power the flipper against the dense medium. Their elbow and wrist joints are fused, so the flipper is rigid and paddle-like and cannot be folded like the wing of a flying bird, and its movement at the shoulder is also more restricted than that of wings used for flight. Penguins' bones are solid, which also helps to reduce their bouyancy and aids diving, and their bodies are streamlined and torpedo-shaped, which is the most appropriate shape for the dynamics of water, reducing the amount of power needed to swim. They use their feet and tail as rudders.

Penguins swim underwater a lot, as this is less demanding than swimming on the surface. They also "porpoise" when swimming long distances, leaving the water briefly like the dolphins, using their impetus to reduce the demand for energy and enabling them to fill their lungs with air. In contrast, the flying auks are a compromise, their body size limited to what their short, pointed wings will support in the air and propel beneath the water surface. Penguins are fast swimmers, able to reach speeds up to eighteen knots for short periods, and can propel themselves several feet out of the water onto the ice. The smaller species like the adelie and chinstrap penguins can dive 300 feet (91 m) and stay underwater for up to six minutes. The emperor penguin dives down 800 feet (245 m) and can stay submerged for eighteen minutes.

Birds' kidneys can only excrete small amounts of sodium chloride, and the excess ingested by seabirds with their food or through drinking seawater could therefore be a problem. To overcome this, the penguins, gulls, cormorants, and tubenoses have enlarged nasal glands, which extract and secrete salt from the blood, and this drips off the end of the bill in the form of a concentrated saline solution. They can therefore eat marine life and drink seawater with impunity, and do not need freshwater. The oil gland is well developed in the penguins, and is located on the rump in the tail feathers. The secreted oil is applied to the feathers to waterproof them, and is most important for birds that spend a lot of time in the water. It is a well-known fact in aviculture that waterfowl that are kept dry with no access to bathing or rainfall can lose their waterproofing, and then become waterlogged and drown when they eventually have access to deep water. It is interesting to speculate how the incubating emperor penguin manages without oil-gland stimulation, when it is not exposed to water or rain during its long midwinter nesting period.

Food

Penguins need cold water for two reasons: to help control their body temperature and to provide their food. It is a common fallacy that warm water contains more animal life than cold water, but unlike the tropical land masses with their enormous number of species and individuals, with the exception of the coral reefs the warm seas are not as rich as the temperate and cold waters. The southern oceans and the currents arising from the West Wind Drift encircling Antarctica and flowing northward are the richest waters in the world, and are home to all the penguins. Warm water flowing south from the sub-Antarctic wells up against Antarctica, bringing nutrients to the surface and nourishing the rich planktonic growth, which is the basis for all life in the sea. The cold, nutrient-laden waters arising from the Antarctic in turn well up against the western coasts of South America and southern Africa, and along the southern coasts of Australia and New Zealand, producing rich feeding grounds for fish and thence for seabirds. In addition, trade winds blowing from the continents roll the water away from the coast, bringing up the cooler and richer water close to shore. Plankton is composed of two basic organisms, plant life or phytoplankton and animal life or zooplankton, and all marine vertebrates ultimately depend upon these tiny organisms, which flourish in massive quantities where the temperature and nutrients are optimum. Phytoplankton is by its very nature restricted to the sea's surface layers, where sunlight converts mineral salts into organic matter through the process of photosynthesis, and it is the first and most important link in the Antarctic food chain. Zooplankton is dependent upon it, and varies in size from microscopic forms to the 2-inch-long crustaceans known as krill, which is the world's most plentiful form of animal life, and the right conditions produce tremendous swarms. One off South Georgia in 1980 covered several square miles and was estimated to weigh 10 million tons. Krill is the main source of food for a number of Antarctic verbetrates from fish to whales. It is eaten by all the Antarctic penguins and is the main diet of the Pygoscelid penguins (the adelie, chinstrap, and gentoo) and the Eudypted species (e.g., Fiordland crested, Snares Island, and erect-crested penguins). Even the penguins that do not eat a lot of krill (such as the fish-eating gentoo penguin and the squid-eating rockhopper, emperor, and king penguins) depend upon food items which are in turn dependent on krill.

An adult blue whale can eat two tons of krill daily, and its depletion together with the other Antarctic baleen whales through overhunting has resulted in increased supplies for the penguins. The amount eaten by penguins is believed to equal that taken by all the seals and whales combined. Studies aboard the ice breaker *General San Martin*, during the 1957/58 Argentine Antarctic Survey of the Weddell Sea, found that krill possessed strong antibacterial properties against aerobic bacteria and upper gastrointestinal tract bacteria. Further studies showed that krill also contained an antibiotic found in the yellow-green algae (*Phaeocytis poucheti*) that they eat, and that the antibiotic could be easily extracted from the plankton with solvents. It was suggested that this natural antibiotic has such an important influence on penguins that they may not survive outside their natural environment without plankton. It is now well known that Antarctic penguins can indeed survive and breed in northern zoos and oceanariums without plankton in their diet, but they are certainly very susceptible to airborne lung infections such as aspergillosis, and elaborate measures are necessary to protect them.

Penguins eat a large variety of other forms of marine life, both vertebrate and invertebrate, in addition to krill. For some species, such as the king, emperor, and rockhopper penguins, squid forms a major percentage of their food; while fish such as pilchards, anchovies, and whiting are important foods for the black-footed, Magellanic, Galapagos, Humboldt, blue, and gentoo penguins. When penguins are fishing some distance from shore they can store food in their crops for several hours until they return to the colony to feed their chicks. Although some fermentation may begin there, the food is not exposed to the potent excretions of the stomach, and like other fish-eating birds the penguin has a large proventriculus or glandular stomach where digestion occurs, and a correspondingly small ventriculus or gizzard, unlike the grain-eating birds that must grind down their seeds.

Breeding

Most penguins are gregarious while at sea and are also highly social during their land-based nesting season, breeding mostly in huge colonies. Those of the adelie penguin on the coasts of continental Antarctica contain several hundred thousand birds, and the colonies of chinstrap penguins on Zavodovski Island in the South Sandwich Islands and of macaroni penguins on South Georgia are each said to contain 5 million birds. On the Falkland Islands there are believed to be at least 2 million rockhopper penguins, and over 1 million royal penguins nest on MacQuarie Island. Where islands offer more than one penguin habitat, such as Beauchene Island in the Falklands, nesting colonies of different penguin species may be adjacent, with rockhopper penguins nesting by the hundreds of thousands

Adelie Penguins   Most penguins nest colonially for safety and because of the lack of suitable sites, especially within the Antarctic Circle. Colonies of the adelie penguin like the one above on the shores of the Antarctic continent may contain several hundred thousand birds.
Photo: Courtesy NOAA

on the lower, level areas and Magellanic penguins nesting in burrows in the tussock grassland on the higher slopes. However, while they are certainly gregarious, to call penguins sociable is rather misleading, for they are actually unsociable, quarrelsome birds, and are very protective of their small territories. Their colonies are also very noisy places, for penguins have an unusual range of calls. The Spheniscid penguins have loud, donkey-like braying calls, the *Aptenoydes* species trumpet, the Fiordland crested penguin makes a deep booming sound, and the blue penguin makes a long drawn-out wail and a dog-like bark. The most peaceful rookeries are those of the foot-nesting penguins—the emperor and king penguins, which incubate their eggs on their feet—for they have no nest to defend.

There are several reasons why penguins nest in such crowded colonies. First, on many oceanic islands and on the shores of the Antarctic continent, the number of accessible beaches and therefore potential nesting sites is limited. Also, unlike most land birds, seabirds do not generally feed close to their nests because they have access to the open sea around them, so they are able to nest in very large colonies in close contact without jeopardizing their immediate food supply. It has also been suggested that there are greater chances of survival because of the greater number of birds watching for predators, and therefore the reduced chances of nest predation. However, warning of approaching danger is mostly of value if the threatened birds can disperse quickly to avoid the predators, which of course the nesting penguins cannot do. Predation on land, by other birds, is virtually limited to the penguins' eggs and chicks, whereas predation in the water of adult or independent juvenile birds is the prerogative of marine mammals, mostly the leopard seal. Colonial nesting may indeed give the individual penguin greater odds against predation of its eggs or chicks, but it certainly does not reduce the overall amount of predation. At isolated nests a skua will dive-bomb and bump a penguin off its nest or entice it away, while its partner steals the egg or chick, but this also happens just as easily at the edges of a huge colony. The larger and more conspicuous the colony, the more predatory marine birds it attracts and supports, so the percentage of predation is probably quite similar. It is certainly safer to nest in the center of a large colony as egg and chick predation there is practically nil, so penguins work their way inward to these favored positions as they age so that the younger breeders on the colony's edges suffer a higher rate of predation. There is an advantage to the colony in this arrangement as the older, established breeders are more likely to have nesting success than birds laying for the first time.

A definite advantage of colony nesting, however, is the penguins' cooperative behavior in the care of the chicks, when they are left in a creche under the care of a few adults while most parents go fishing, thus providing more food and increasing the chicks' chances of survival when they eventually head to the water themselves. However, this may not reduce chick predation, and may in fact increase it, as skuas are especially dangerous at creches. They panic the chicks and then pick them off as they disperse, while the few adult chaperones run around frantically trying to protect them. The penguins that nest in burrows are definitely safer from predation than those that lay their eggs in the open. With the exception of the emperor penguin, which makes a long trek over the sea ice to nest, and some of the crested penguins, which lay their eggs in the forest some distance from the shore, most species nest on the beach or in the adjoining sand dunes or tussock-covered hillsides close to the sea. They usually keep the same mates for several years, up to eleven years in the case of the blue penguin. The yellow-eyed, black-footed, and blue penguins breed when only two years old; the gentoo, adelie, emperor, and king penguins when they are three to four years of age, but the royal penguin is not sexually mature until it is five years old. The blue penguin and the Galapagos penguin are occasionally double-brooded, but penguins normally lay only a single clutch each year, although some of the more northerly species may take advantage of the longer summers to lay again if their first clutch is lost. Penguins of the genus *Aptenoydes*—the foot-nesting emperor and king penguins—lay a single egg, but all other species lay two and occasionally three eggs. In these species, especially the rockhopper, Snares Island, macaroni, and Fiordland crested penguins, there is considerable size difference between the eggs, the second being much larger, up to

70 percent bigger in the macaroni penguin. In each case the smaller nestling dies within a few days, strange behavior that ensures that at least one chick is raised in each nest. If only one egg was laid, it could be lost as a result of predation or aggression between nesting neighbors. If both eggs hatch, only the larger chick survives, whereas if they were of equal size and both received equal amounts of food the effort could then be totally wasted if the parents were unable to meet the demands of both chicks as they matured.

Penguins have several types of basic nesting behavior. The king and emperor penguins incubate their single egg on their feet covered by a flap of their stomach skin. The three species of the genus *Pygoscelis*—the adelie, chinstrap, and gentoo penguins—lay their eggs in an open scrape among the pebbles or sparse soil, generally because there is not the depth of soil to do otherwise. The six species of crested penguins of the genus *Eudyptes* nest either colonially in the open or within a few hundred yards of the shoreline, with the exception of the Fiordland crested penguin, which nests in dense, wet forest. All the other penguins dig burrows in the sand or peat, or occupy a crevice between rocks, under a shrub, or in a hollow under the roots of a tree, although the black-footed penguin also nests in large surface colonies in the open. The eggs of the penguins that nest on pebbles have shells 50 percent thicker than those that lay on the sand or soil in burrows. The incubation period for the eggs of the small and medium-sized penguins is between thirty-five and forty-two days, but the larger species take considerably longer. The sixty-four days needed to hatch the emperor penguin's egg is one of the longest incubation periods of all birds, although still fourteen days short of the average for a kiwi's egg.

Penguins, especially the Antarctic species, suffer high chick mortality. They may be chilled in exceptionally cold weather, or receive insufficient food from the parents when heavy pack ice prevents them from returning frequently to the nest to feed their chicks or the brooding parent, causing it to desert the nest. If the sea is rough when the chicks swim out for the first time they may be so buffeted by the waves that they are unable to feed, and the mortality rate is very high. Also, their lack of experience makes them especially vulnerable to leopard seals, and it is believed that only about 20 percent of emperor penguin chicks survive their first year of independence. Young blue penguins are also known to have a hard time when they first go to sea, and many bodies are washed ashore when their chance at independence coincides with heavy seas.

Predation

Penguins lost their flight and survived for many years in the total absence of mammalian land predators, which was very noticeable when they encountered the first Antarctic explorers and their dogs, regarding neither with alarm, and were easily killed for food. Their reaction to danger occurs only in the water, the home of their traditional mammalian predators—the leopard seal and the killer whale. However, although there are no mammalian predators on the Antarctic continent and oceanic island breeding colonies, nesting penguins are harrassed and preyed upon by several species of flying seabirds. Sheathbills are major predators. Unusual white birds about the size of a small gull, but with un-webbed toes, they are the connecting link between the waders and the gulls, and have a stout bill with a horny sheath covering the nasal openings. They are scavengers at Antarctic and sub-Antarctic penguin colonies, eating carrion or startling the returning adult penguins into regurgitating the food they have brought back to the nest; but they are also adept at stealing eggs and young chicks. The other major predators are the skuas: large gull-like, piratical seabirds, with sharply hooked upper mandibles. In the absence of penguins they are mostly scavengers, harrassing flying birds up to the size of a gannet, chasing them and seizing their wings or tail until they disgorge their latest meal. They are also distinctly predatory and eat the young of other skuas that wander into their territory, and on MacQuarie Island they kill the introduced rabbits and rats. Skuas take a great toll at the penguin rookeries, where they usually confine their activities to the fringes of a colony and often work in pairs, one bird distracting the nesting penguin while the other steals its egg or chick. They are especially dangerous at creches where they scatter the poorly protected young and then pick them off with ease. Colonies of a million or more penguins obviously attract and support large numbers of predatory seabirds. The Antarctic skua, one of the most common species and a wide-ranging seabird of the southern oceans, is probably the worst penguin predator. On South Georgia, where it preys heavily on gentoo penguins, observations have shown that each pair of skuas has sole access to the penguins living in its territory. On Gough Island, it was estimated in 1956 that there were almost 3,000 breeding pairs of Tristan skuas, which were the major predators at the rockhopper penguin colonies. The number of skuas can rise dramatically when an unnatural source of food suddenly becomes available to them. Those living near the French Antarctic base of Point Geologie in Adelie Land, which ate carrion and adelie penguin eggs, increased from about 90 in 1966 to over 300 within 4 years due to the plentiful supply of food at the base's garbage dump. Their egg-laying season increased and most pairs were able to raise two chicks per nest.

Predatory land birds are also a problem for the penguins nesting on the South American coast and the islands off its southern tip, including the Falkland Islands. Several species of birds of prey there are known to take penguins and their eggs. The caracaras, an endemic group of New World raptors, are opportunistic feeders, scavenging around penguin colonies and taking eggs and chicks when the opportunity arises. Forster's caracaras (*Phalcoboenas australis*), of Tierra del Fuego and the Falkland Islands, have gorged so much at penguin colonies they could not get airborne. The turkey vulture (*Cathartes aura*), a wide-ranging neotropical scavenger, is a predator of nestling herons and ibis, and of penguins' eggs and chicks. The Peruvian gull and the kelp gull are also penguin predators in Peru and Chile.

Land mammals are also a hazard to penguins nesting on the South American mainland, where the colpeo (a type of fox) and domestic cats and dogs prey upon them; but the most incongruous mammalian predator of penguins must be the leopard, which raided nesting black-footed penguins on South Africa's mainland coast. The most serious predation of penguins by land mammals occurs in New Zealand, where dogs, cats, rats, ferrets, stoats, and weasels have access to the nest sites of all penguins nesting on the mainland—the blue, Fiordland crested, yellow-eyed, and white-flippered penguins. Dogs have been particularly damaging to penguin populations, killing over fifty blue penguins at Oamaru colony in 2001, and the colony at Piha Beach, Auckland, was wiped out by dogs in 1982. On Isabela, one of the two islands on which the Galapagos penguin breeds, introduced rats and feral cats, dogs, and pigs all threaten the nesting birds. The narrow Bolivar Channel, which separates Isabela from neighboring Fernandina where the penguin also occurs, has so far proved a barrier to the predators.

Habitat damage and loss have also severely affected some penguin populations. In the Falkland Islands the burning of the tussock grass killed many nesting penguins, and the introduction of domestic grazing animals reduced their nesting habitat. On Campbell Island, a major home of the rockhopper penguin in Australasia, the government destroyed 900 feral sheep in 1990, clearing it for the first time since they were introduced in 1895. The damage they caused to the tussock grass was believed to have contributed to the drastic decline in penguin numbers. At sea, the leopard seal is the penguin's major predator in Antarctic waters. The seals, which grow to 11 feet (3.3 m) long, lurk at the edge of the ice pack waiting for the penguins, which leap up to 6 feet (1.8 m) onto the ice shelf in their frantic efforts to avoid them. When a seal seizes a penguin, it first bites off its feet and tail and then beats it on the water surface to skin it. Killer whales kill penguins in Antarctica and along the coasts of South America, where sea lions have also been observed chasing penguins.

Penguin predation by humans did not reach threatening proportions until historic times. Nesting penguins were certainly gathered in southern South America, where the Tierra del Fuego Indians ate them and their eggs and made clothing from their skins. Penguins also coexisted with the early hunting and gathering communities that inhabited New Zealand and southern Australia. This situation changed with the arrival of Europeans in penguin nesting areas. Penguins were gathered in large numbers for their meat and skins, and for boiling for their oil, and whole colonies were wiped out through excessive egg harvesting. Settlers destroyed their habitat through land clearance and farming, burning their tussock-grass nesting habitat together with many incubating penguins, their eggs, and chicks. They introduced grazing animals that destroyed more penguin habitat, and carnivores that preyed on the birds and made nesting even more hazardous.

In the late nineteenth century the densely feathered penguin skins were in great demand, especially for trimming clothing, and rockhopper penguins were killed in large numbers on Tristan da Cunha for their head plumes. Their heads were skinned and the "pelts" dried and used for decorating caps and bags or were sewn into mats which contained several dozen skins. Moulting penguins were rounded up and plucked of their loose feathers for stuffing mattresses and cushions. They also proved to be a surprisingly rich source of oil, and were used heavily for this purpose, especially when the stocks of marine mammals were depleted by the sealers. The king penguin, weighing 35 pounds (16 kg), was first used as fuel for rendering the blubber of marine mammals but by the end of the last century the sealers had exterminated the fur seals and elephant seals on some islands, and turned their attention in earnest to the penguins. Their fat produced good quality lamp oil and they were boiled at the rate of 400,000 annually, which number apparently produced 50,000 gallons of oil. Rookeries were either severely depleted or totally wiped out. One huge colony, which covered 40 acres (16 ha) in 1815, was eliminated by the end of the century. On MacQuarie Island the king penguin was nonmigratory so the colonies were not restocked through migration, and their harvesting became uneconomical when the total island population had dropped to just 6,000 birds in 1913; the sealers then turned their attentions to the royal penguin. A much smaller bird, weighing only 10 pounds (4.5 kg), it was an endemic species, nesting nowhere else, and was killed at the rate of 150,000 annually. The resultant public outcry in Australia fortunately ended the slaughter before the species was exterminated. MacQuarie Island has been protected since 1933, and only the rusting machinery and boat hulks remain as a reminder of the horrendous industry. Only a few thousand king penguins live there now, although almost 2 million rockhopper and royal penguins also nest there. The Falkland Islands' penguins also suffered severely at the hands of the blubber harvesters. In the early years of settlement several hundred thousand rockhopper and king penguins were killed annually. The king penguin colonies were easy to pillage as they were situated on flat, open country, and the birds were rounded up and clubbed. They were soon eliminated, although a few have recently returned.

None of the sub-Antarctic islands were safe from plunder. Early in the twentieth century, gentoo, king, and macaroni penguins were slaughtered at the rate of 500,000 annually on South Georgia by the elephant seal hunters, who boiled them for their oil and used them to stoke the fires that rendered the seal blubber. All three species survived the onslaught and still occur there. On remote Heard Island, Australian territory in the southern Indian Ocean, uncontrolled hunting between 1855 and 1929 reduced both the seal and penguin populations to such low numbers that visits ceased in 1930 because they were no longer commercially viable. Fortunately, the neighboring McDonald Islands did not suffer, because of their inaccessibility, and served as a reservoir for the natural restocking of Heard Island after the era of the sealers. The first landing on the McDonald Islands, by helicopter, did not occur until 1971, and they are the only sub-Antarctic island groups free of introduced plant and animal pests. Their value and need for protection has been recognized by UNESCO'S Man and Biosphere Program and the World Conservation Union.

Threats

Egg collecting was also a very serious threat to penguins in the nineteenth century and well into the twentieth century, and occurred on such a vast scale that whole colonies disappeared. Dassen Island, off the coast of South Africa, was the major breeding site for the black-footed penguin, and between 2 and 3 million penguins nested there early in the twentieth century. Several hundred thousand eggs were collected annually between 1917 and 1947, by which year the penguins had been reduced to 100,000 birds, yet egg collecting continued until 1968. In the Falkland Islands 85,000 rockhopper penguin eggs were collected annually on

100-acre (40 ha) Kidney Island alone, and by 1952 so few penguins survived that only 1,000 eggs could be gathered.

The drastic decline of the once-plentiful Humboldt penguin to its present endangered status began in 1850 with the commercial harvesting of guano—the accumulated droppings of millions of seabirds over many years. It was hastened more recently by overfishing and by the natural change in climate caused by the El Niño phenomenon. In the 1920s, 90 million seabirds—cormorants, boobies, pelicans, and penguins—were estimated to nest on Peru's offshore islands. This region has one of the world's driest climates, and in the absence of rainfall the guano had accumulated for several thousand years and in places was almost 150 feet (45 m) deep. Between 1850 and the 1920s it was harvested commercially, disrupting the penguins' breeding activities. Their surface breeding sites were soon destroyed and eventually deep guano was no longer available for them to dig their nest burrows. The gangs of Chinese coolies who dug and loaded the guano onto vessels also collected adult penguins and their eggs for food. Although large-scale guano digging ended years ago, it was still collected year-round until 1986, when the Peruvian government finally agreed that it should only be gathered outside the penguins' breeding season, but the penguins were still disturbed by the guano collectors. Finally, realizing the value of this natural resource, the government protected the birds that produced the guano, but it did not protect the birds' food supplies. When the commercial value of the vast shoals of anchovies that nourished the millions of seabirds was recognized in the early 1950s, Peru's fishing fleet escalated, and the collapse of the guano industry from overharvesting was quickly followed by the rise of the anchovy fishery. Vessels hauled aboard 8,863,000 tons of fish in 1964, according to FAO World Fishery Statistics, 15 percent of all the fish caught in the world that year. By comparison the catch in 1938 had been only 23,000 tons. Overfishing had the obvious result and by the late 1970s the catch was down to 1 million tons annually. For the birds, official protection was not enough, and millions died when deprived of their food supplies. Many were also accidentally caught in the fish nets, and like the coolies a century earlier, the fishermen raided the nesting colonies for birds and eggs.

The threats to penguins may have changed in recent years, but they are still numerous. Originally they were more direct, and involved killing them for food or oil and collecting their eggs, and the regional threats of guano collection and overfishing. Nowadays the threats to penguins are mostly indirect ones, but they are more international in their scope, and are likely to have more serious repercussions on the present-day penguin populations, several of which are seriously reduced. Like the overfishing of Peru's anchovies thirty years ago, competition for fish is still one of the most serious threats to the survival of penguins, but is now more wide-ranging and affects far more species. Commercial fishing began in Antarctica in the mid-1960s and the fish stocks there have since been heavily exploited. Within fifteen years *Notothenia rossii*, the dominant fish in South Georgian waters, had been reduced to one-tenth of its former numbers. Now the commercial fisheries have discovered krill, and trawlers, mostly of Japanese and Russian origin, harvest several thousand tons annually, a catch that is expected to increase dramatically. Overfishing by the European and Asian fleets was cited as a major reason for huge penguin losses in the Falkland Islands in the mid-1980s, when the vessels operated within three miles of the shore because of the British government's unwillingness to enforce a 200-mile no-fishing zone. Their annual catch was estimated at 1 million tons, and the Asian vessels also took large quantities of squid, an important item of the rockhopper penguin's diet. Reduced to a fraction of their earlier numbers by egg collectors, the black-footed penguins off South Africa's west coast declined further following the overexploitation of their favorite foods—pilchards and anchovies—but the pelagic fish industry there fortunately collapsed in the mid-1980s before the penguins succumbed.

The exploration and scientific study of Antarctica and the sub-Antarctic islands has been detrimental to the penguins. Competition between them and people for the best beaches has adversely affected penguin populations, especially in Antarctica, where the establishment of so many national scientific stations severely disrupted colonies. Both penguins and humans naturally select the most accessible beaches, either for nesting or access for base and landing strip construction and operation, and a number of stations have been built on traditional penguin nesting beaches. When Hallett Station was built in 1957 a rookery of 8,000 adelie penguins was disrupted, and a later extension of the base displaced many more.

Oil spills are also a major threat to penguin survival, especially black-footed penguins in the busy sea-lanes off the tip of South Africa. Hundreds of vessels round the Cape of Good Hope every week, many of them supertankers carrying thousands of tons of crude oil. Several accidents there have resulted in major oil spills, which have killed many penguins. The latest in three decades of serious spills occurred on 29 June 1994, when the *Apollo Sea*, carrying 25,000 tons of crude, hit rocks and sank between Robben and Dassen Islands. Almost 10,000 oiled penguins were rescued and treated by the South African National Foundation for the Conservation of Coastal Birds, but for every oiled bird caught on the beach it is thought that at least ten died at sea. Only 4,300 of the rescued birds recovered and were eventually released, but sightings were made of almost half of these the following year and many bred successfully. Compounding its effect on the penguins, this spill occurred at the height of the main nesting season, and over 6,000 chicks were estimated to have died at the rookery because of the loss of their parents. The chicks can survive, however, if only one parent is left to care for them. Two small chicks were left in the nest when an oiled bird was removed from Robben Island. One month later it was returned to the island after rehabilitation, and within a day found its nest, which contained two large chicks its mate had raised single-handedly. Prior to this disaster the two islands were home to 40,000 penguins, which was believed to be almost one quarter of the species total.

It is not only the major sea routes that are becoming hazardous to penguins and other seabirds. Oil spills have already occurred in remote places like the Galapagos Islands and Antarctica, especially as a result of the great increase in ecotourism. In 1989 an Argentine tourist vessel sank near the Litchfield Protected Area of the western Antarctic Peninsula, releasing 125,000 gallons of diesel fuel. To make matters worse, krill leaping from the water to escape the oil attracted seabirds to it. The Galapagos Islands have also already suffered their first oil spill; when the Ecuadorian supply ship *Iguana* discharged 50,000 gallons of diesel fuel into

Black-footed or Jackass Penguins *Oil spills are a constant hazard to penguins, especially those living on the South African coast, like the jackass penguins above on their nesting beach near the Cape of Good Hope. Having suffered severely from egg collecting during the first half of the twentieth century, which considerably reduced their populations, they must now contend with supertankers attempting to negotiate the "Cape of Storms."*
Photo: Clive Roots

Academy Bay, Santa Cruz, after hitting a reef in 1988. This devastated the marine invertebrate life there, including the squid and octopus, which are important items in the diet of the flightless Galapagos cormorant and Galapagos penguin.

High penguin mortality in the Falkland Islands in the summer of 1985–1986 caused worldwide concern. Three thousand adult rockhopper penguins died in a colony on New Island and dead rockhopper and gentoo penguins were washed ashore on many other islands. Magellanic penguins also died in large numbers but were less visible due to their burrow-nesting habits. The deaths coincided with the observation that there was a high level of infertility in rockhopper penguin eggs. The many opinions offered for the deaths and infertility included overfishing and water temperature fluctuations, which affected the penguins' food supplies. Birds autopsied by the British Ministry of Agriculture were found to have died of starvation, and had high concentrations of lead in their liver and kidneys, which could have so weakened them that they were unable to feed. Shortage of food resulting from commercial overfishing could have been a major factor, for European and Asian fleets were very active in the area in the early 1980s. However, high mortality was also observed in the Magellanic penguin colonies on Argentina's east coast; on

South Africa's offshore islands the black-footed penguins had a very poor breeding season, and deaths of both young and adult yellow-eyed penguins occurred on New Zealand's south coast. It was believed that these widespread breeding failures and high mortality rates may have been caused by oceanic temperature fluctuations around the Southern Ocean's northern margins, resulting in warmer surface water. This was supported by the observation that many northern warm-water seabirds flocked to the Falkland Islands in the summer of 1985/86, including species never before recorded there. It was therefore thought that the irregular, naturally occurring phenomenon known as the "El Niño" may have been responsible for the oceanic temperature increases. This regional event occurs periodically when warm water from the equatorial region enters the eastern Pacific Ocean, increasing its water temperature by several degrees. The warming of the surface water reduces the krill and then the fish populations, and the seabirds' numbers drop dramatically. It may also prevent the vertical migration of cuttlefish and octopus to within diving range of the penguins and in the eastern Pacific Ocean has periodically had a serious effect on the Galapagos and Humboldt penguins. The Galapagos penguin is especially vulnerable to the fickle nature of the ocean environment because of its very restricted range and exposure to alien predators, at least on one of its two islands. Its numbers dropped from over 1,700 in 1980 to 400 in 1983 as a result of the 1982/83 El Niño, because it could not disperse from food-depleted areas as rapidly as flying seabirds. However, even after these dramatic losses the penguins had recovered to their earlier numbers by 1986.

In 1982/83, krill in the southern oceans declined dramatically, according to the Scripps Institute of Oceanography and other research vessels operating there. The water temperature was 37.4°F (3°C), warmer than normal by two or three degrees, and the El Niño of 1982/83 was believed responsible. Warming of the surface water prevents the upwelling of the lower colder and richer layers, which bring nutrients to the surface-dwelling plankton. The warmer waters do not have the rich planktonic flora and fauna that is present in temperate waters and so abundant in cold, sub-Antarctic seas. The entire population of flying seabirds (at least 10 million) disappeared from Christmas Island in the central Pacific Ocean after the 1982/83 El Niño, but most species returned and the populations recovered in a few years.

The results of the occasional El Niños in the eastern central Pacific Ocean around the Galapagos Islands, and possibly now even farther south around the Southern Ocean's northern margins, have shown how devastating the warming of the seawater can be to the food chain, and eventually to seabird populations, resulting in breeding failures and high adult mortality. This raises the specter of the disastrous effects likely from the warming of the world's oceans on a larger scale. The burning of fossil fuels releases millions of tons of carbon into the atmosphere annually, which traps heat to produce the greenhouse effect and global warming. Unfortunately, the world's carbon emissions are expected to increase dramatically as developing nations expand their fossil fuel plants, despite the fact that the developed nations have offered to share their technology with the Third World, while being generally unwilling to place reduction targets on their own emissions. Increased global warming could melt the polar ice caps, warm the Antarctic waters, raise sea levels, and alter climatic conditions worldwide. It is the most serious threat to the penguins and other animals of the cold southern waters, and its effect on those creatures would undoubtedly make the El Niño consequences appear trivial.

There have already been many indications that global warming is creating problems in the Southern Oceans. When yellow-eyed penguins continued to decline on Campbell Island early in the 1990s, water temperature change and the depletion of fish were cited by New Zealand's Department of Conservation as possible causes. Surveys and studies similarly showed a decline in the rockhopper penguin colonies on Campbell Island, for which increased water temperature and reduced fish stocks were also cited. Many adult blue penguins died in Bass Strait in 1985, and 80 percent of the chicks died that year in the colony at Phillip Island's Penguin Parade Reserve. These losses were all attributed to starvation, and although the reason for the food shortages could not be determined, increased water temperature was suspected.

The Species

Antarctic Penguins

The present-day Antarctic continent, the nucleus of Gondwanaland, became isolated and surrounded by water as the continents drifted apart about 100 million years ago. Winds caused by earth's rotation created the circumpolar current or west wind drift, with air masses circling the continent in a clockwise direction. This isolated Antarctica from the warmer northern temperatures, and produced the conditions that created the great ice sheet that now covers most of the land mass. Accumulating for over 20,000 years, it is now about 2 miles (3.2 km) thick at the South Pole and is so heavy that the bedrock has been depressed 2,000 feet (600 m). More than 90 percent of the world's ice now covers Antarctica, a land twice the size of Europe. The Antarctic continent includes the great ice cap and its coastal zones; the Scott, Balleny, and Peter I Islands, and most of the Antarctic Peninsula, the tip of which is the only part of the land mass outside the Antarctic Circle. The water temperature of the Antarctic seas adjoining the ice pack is about 28°F (−1.8°C) for much of the year, the temperature at which salt water freezes.

All of Antarctica's land is permanently ice covered except for about 3,000 square miles (4,800 square km) of rock, shingle beaches, and soil around the continent's edge and on the Antarctic Peninsula. The ice sheet reflects most of the solar radiation back into space, giving the Antarctic a colder climate than the Arctic, where some of the ice melts each summer, allowing the open water to absorb the sun's heat. As it is the difference between the equatorial and polar temperatures that powers atmospheric circulation, winds are stronger and more frequent in the south, and winds up to 155 mph (250 kph) have been recorded. It is an ecosystem unlike any other, and in winter is the coldest place on earth, with a world record low temperature of −128°F (−89°C). In addition to these extreme conditions the long winter night blankets the continent in darkness for four months.

Emperor Penguins *Confined to Antarctica, the emperor penguin breeds in the middle of winter (August), the males incubating the single egg on their feet covered by a flap of their stomach skin. During the summer, which begins in October, the growing chicks congregate in creches where they are supervised by an adult, allowing the other parents to go fishing.*
Photo: Courtesy Harcourt Index

There are no resident land birds on the Antarctic continent. The absence of plant and animal food, the low temperatures, and the exposure factor have prevented colonization of the continent's plateau. Many seabirds utilize the milder coastline, where summer temperatures are at or just above freezing and the rich adjacent seas provide ample food. Three species of penguins breed within the Antarctic Circle. The emperor penguin nests solely there, venturing some distance inland to lay its single egg in midwinter, and seldom swims far from the continent's coastline, even in summer. The adelie penguin is also a true Antarctic species, nesting on the edges of the continental land mass and on islands within the Antarctic Circle, although its range does extend to the sub-Antarctic islands. A third species, the chinstrap penguin, nests on a few islands within the Antarctic Circle, but the bulk of its range is outside it on the Antarctic Peninsula and on many islands farther north, so it is not a truly Antarctic bird. Both Antarctic penguins nest in the open, the emperor penguin on the ice, and the adelie penguin on the shingle beaches or on the ice pack, as there are no opportunities for digging burrows. During the short Austral summers, from October to February, the temperature hovers around the freezing point, and even where the snow and ice have melted it is still a very barren land, the sparse plant growth being mostly mosses and lichens.

Emperor Penguin (*Aptenodytes forsteri*)

The emperor penguin is the world's largest penguin, and the largest flightless waterbird, reaching a weight of 65 pounds (30 kg) and measuring 45 inches (1.14 m) from the tip of its bill to the end of its rump. Its body is pale bluish-gray with a black border extending from the sides of the neck to the flanks, and with white underparts that have a yellowish tinge on the upper breast. Its head is blackish-blue with large yellow-and-white auricular patches, and its long bill is also dark with a thin red line along the bottom mandible from the gape almost to the tip. Its legs and feet are dark gray. Males and females are alike externally, and there are no subspecies. The emperor penguin is confined to the Antarctic continent, where it feeds in the waters immediately adjacent to the land mass. It rarely strays beyond the waters of the Antarctic Circle, but there have been recent sightings of juvenile birds off the coast of Argentina. The emperor penguin dives down to 800 feet (245 m) and can stay submerged for eighteen minutes, receiving oxygen from the oxyhemoglobin and the oxymyoglobin stored in its muscles. Several pounds of stones have been found in its crop, giving rise to the belief that this may act as ballast for deep diving, since "grit" for grinding down its food is hardly necessary for a fish-eating bird.

Despite the harshness of the climate the emperor penguin breeds in midwinter, unlike all other species, simply because the summers are too short for its lengthy courtship, egg laying, incubation, and chick raising, which require at least five months to complete. As winter closes in at the end of March the penguins have difficulty breaking the ice and seals take over their dive holes, blocking their access to the sea. They then come out onto the ice shelf to breed, nesting mainly on the coasts of Victoria Land, Wilkes Land, and Enderby Land, with a total of about thirty known colonies. The largest colony is on Coulman Island, where 25,000 pairs gather to lay their eggs. In 1994 a new colony of several thousand birds was discovered on the Budd Coast of Wilkes Land, and it is believed that other unknown colonies may survive on Antarctica's relatively unexplored eastern coast. However, they are not all content to nest close to the frozen ocean and many make long treks inland, often traveling up to 200 miles (320 km), to Cape Crozier, on the Ross Ice Shelf, which is their most southerly breeding site. Captain Scott's team, led by Dr. Edward Wilson, took nineteen days to reach Cape Crozier, and lived there in a stone hut heated by a stove fuelled with penguin blubber.

The penguins return to their colonies in March and April, tobogganing over the ice on their bellies. At distant Cape Crozier they nest in May, a month later than those at sites farther north. The last day of daylight in Antarctica is April 25, so courtship, egg laying, and incubation all take place in total darkness during the long winter night, which lasts four months. When they have laid their eggs the females pass them over to the males, who often need a lot of nudging to encourage them, and then head back to the open sea. The males alone incubate the eggs, fasting all the time and eating snow for moisture, with the result that they are 45 percent lighter at the end of the nesting season. To conserve heat their metabolic rate drops and they become very sluggish, and several thousand incubating males huddle together, packed ten birds to a square yard (.83 square m), and changing places regularly so that each gets a turn in the middle. This helps to conserve at least half the body heat they would otherwise lose to the extreme low temperatures and tremendous blizzards. Even so, in such extreme conditions their stored body fat barely lasts until incubation is completed, especially as they must be prepared to feed their hatchlings with an oily crop secretion until the females return. Despite their single egg being the smallest egg of any bird in relation to the adult's size, it is still a major feat of temperature maintenance to hatch it in such extremely cold conditions. They must maintain an incubation temperature of 86°F (30°C), balancing the egg on the feet where arteries keep it warm, and at all times covering it with a flap of their belly skin. The males move away from their incubating huddle when the eggs begin to hatch in August, which is the coldest month, when temperatures drop to −67°F (−55°C). The females are then beginning to return to the colony with their crops full of squid, and will take over the nest duties from the males, who immediately begin the long trek to the sea to replenish their energy. As the sea ice has by this time doubled the size of Antarctica, the males have an even longer journey back to open water, unless by chance the ice has cracked open near the rookery. While they are away, the females feed the young with regurgitated food. When the males return, replenished, both parents share the duties of raising their chick, bringing freshly caught food more frequently as the sea ice begins to break up. As the baby penguins grow they congregate in nurseries or creches, where they are supervised by a few older birds. The adults crowd around to shelter them during blizzards, but cannot prevent them from being covered with snow and ice. The four-month-long summer begins in early October and the adult penguins then take to the water as the ice breaks up. The chicks are left to fend for themselves, although they will not be fully fledged until December, when they enter the water for the first time. It is a harsh world for inexperienced juvenile penguins, and 80 percent are believed to perish in their first year at sea. (See the color insert.)

Adelie Penguin (*Pygoscelis adeliae*)

The adelie penguin is the only other truly Antarctic species, a small bird in comparison to the emperor penguin, measuring about 26 inches (66 cm) long, and weighing 9 pounds (4 kg). It is one of the stiff-tailed or brush-tailed penguins, with a long tail that drags the ground when it walks. Its upperparts and head are bluish-black and the underparts white, and its distinctive white iris gives it a button-eyed appearance. Its stubby bill is reddish-black with a black tip, and its legs and feet are pink. The sexes are alike externally, and there are no subspecies. This penguin is not confined to the polar land mass, although its largest concentrations are within the Antarctic Circle on the continent's coasts and islands, where it is the commonest penguin, with colonies of up to 1 million birds. It also nests on the Antarctic Peninsula, and on the South Orkney, South Sandwich, South Shetland, Bouvet, and MacQuarie Islands; and has wandered as far as Kerguelen and the Falkland Islands.

Adelie penguin numbers have multiplied in recent years at some colonies, possibly due to the increase in krill resulting from the overkill of the plankton-eating whales. Unlike the emperor penguin, it leaves the water in spring to mate and nest, beginning to breed earlier than the other Pygoscelid species (the chinstrap and gentoo penguins) because of Antarctica's shorter summers. The adelie penguin incubates its eggs in the normal bird fashion rather than on its feet like the emperor penguin, so it is not restricted to a single egg and normally lays two, in a slight hollow or on a small pile of pebbles to keep them from rolling away. Even so, when they leave the sea in October they may have to trek 15 miles (24 km) over the sea ice to their nesting colony on the shore. Their most southern nesting site is also at Cape Crozier, which is mostly stones and gravel during their breeding season, unlike its harsh conditions when the emperor penguins nested there a few months earlier. They normally have the same mate and use the same nest site each year, although the female does not accept the male until he has collected stones for the nest. Both birds then share the thirty-six-day incubation duties and take turns returning to the sea to catch krill, which they feed to their young by regurgitation. They have the unusual habit of forcing their chicks to chase them through the colony and catch them before they will feed them.

Even nesting during the Antarctic summer can be a risky business, however. Unseasonably cold weather sometimes prevents the breakup of the sea ice, forcing the hungry penguins to abandon their nests and return to the distant open sea to feed. Blizzards also kill many nesting birds and their young. Adelie penguin chicks fledge when they are about fifty days old, and by March most are ready to take to the sea and head north for the winter. Skuas eat the chicks that have not grown sufficiently to make the journey. Several aspects of the Adelie penguin's behavior are noteworthy. They have a unique display, calling with their head pointed upward and their body rigid, while rapidly beating their wings. Also, they leap several feet out of the water onto the ice at great speed, especially if a leopard seal has been seen nearby; and they have unusual moulting habits, where they shed practically all their feathers at once and take three weeks to replace them, during which time they lose almost half of their body weight.

Maritime and Sub-Antarctic Penguins

The Maritime Antarctic and the sub-Antarctic islands are home to five species of penguins—the king, gentoo, macaroni, chinstrap, and royal penguins. The Maritime Antarctic includes the west coast of the Antarctic Peninsula and its off-shore islands, plus the South Orkney, South Sandwich, and South Shetland Islands. It also includes more northerly Bouvet Island, which is often surrounded by pack ice from the Weddell Sea and consequently is shrouded in fog, despite being on the same latitude as the Falkland Islands. Temperatures in the Maritime Antarctic are milder than on the continental plateau, and summers are wet, windy, and usually close to freezing. The vegetation is limited to mosses and lichens, and there are no burrowing opportunities for nesting penguins.

The sub-Antarctic islands lie in the cold southern oceans, mostly between the continent and the Antarctic Convergence, which lies about 1,000 miles (1,600 km) offshore. This is where the cold surface water flowing north meets the warmer sub-Antarctic water flowing south, which is eventually forced upward by the land mass, bringing nutrients to the surface. These islands include South Georgia, Heard, Kerguelen, and MacQuarie, plus others farther north such as Crozet and Prince Edward, whose climate is affected by the sinking, north-flowing cold Antarctic water that wells up against them. They have a milder climate than the Maritime Antarctic, with snow in winter, but summers last half the year with temperatures ranging from freezing to about 46°F (8°C). Mist and cloud frequently cover the islands and the high precipitation encourages low plant growth which is dominated by tussock grass with some sedges and leafy herbs. The thick peat deposits that nourish this vegetation produce bogs with ponds and drier areas for the nesting penguins.

The open ocean between the ice cap and the Antarctic Convergence covers approximately 9 million square miles (23 million square km) in midsummer, an area equivalent to the combined land masses of South America and Australia. These waters are rich in animal life, with forty species of seabirds occurring there, including seven species of penguins, as the two Antarctic species fish there also. In contrast, the only resident land bird, and the most southerly songbird, of the islands lying within the Antarctic Convergence is the endemic South Georgia pipit (*Anthus antarcticus*), which lives along the shoreline in winter, feeding on tiny crustaceans and molluscs, and ventures inland to nest during the summer months. On milder MacQuarie Island, at the edge of the Antarctic Convergence, the only resident land birds are the starling (*Sturnus vulgaris*), redpoll (*Carduelis flammea*), and flightless weka (*Gallirallus australis*), all aliens. The island's endemic land birds were the flightless rail (*Hypotaenidia macquariensis*) and the parakeet (*Cyanorhamphus novaezeelandiae erythrotis*), both now extinct.

King Penguin (*Aptenodytes patagonica*)

This is the second largest of the living penguins, standing 36 inches (92 cm) high and weighing 35 pounds (16 kg). Its upper body is silvery-gray with a black border extending from the throat down to the flanks. Its underparts are white, tinged with yellow on the upper breast, and it has bright-orange auricular patches in a closed teardrop shape on its otherwise blackish-brown head. The similar but larger emperor penguin has yellow and white earpatches in an open teardrop shape. The king penguin's legs and feet are black, and its top mandible is also black although the bottom one is yellowish-orange with a black tip. In outward appearances the sexes are alike.

Although circumpolar in distribution, occurring on many islands in the southern seas, the king penguin lives mainly on the sub-Antarctic islands between 46°S and 55°S, east of South America's tip—the Falklands, Staten, and South Georgia—and those bordering the Antarctic Convergence in the South Indian Ocean—Prince

Edward, Crozet, Kergeulen, Heard, and the McDonald Islands. Few still nest on MacQuarie Island south of New Zealand, never having recovered their former numbers after the massive slaughter for their oil. The current population of the species is believed to be about 1 million birds, with the largest concentrations on Prince Edward and Crozet Islands. Two subspecies are recognized: *A. p. patagonicus*, which breeds on South Georgia and the Falkland Islands, and *A. p. halli*, which breeds on Marion, Crozet, Kerguelen, Prince Edward, Heard, and MacQuarie Islands. King penguins have wandered north to South Africa and southern Australia.

The king penguin is a social bird that nests colonially, with the colonies varying in size from a few pairs to several thousand. Preferred sites for a nesting colony are the barren shores of the sub-Antarctic islands, moraine and beachside valleys free of snow, and among the tussock grass farther inland. Like the emperor penguin, it is a foot-nester, incubating its single egg on top of its feet, protected and warmed by a fold of the stomach skin. Although it has the opportunity to burrow into the peat moss of the tussock grasslands, it is believed that its size prevents this. It is a summer nester, and begins to return to its breeding colonies in September, although late breeders do not arrive until November. The single egg is laid between October and March every other year, and the incubation period is fifty-five days. The male begins the incubation and does so solely for three weeks while the female is at sea feeding, and she then returns to relieve him. When the egg hatches both parents share the chick-brooding and food-gathering duties until it is five weeks old, when it joins a creche with many other chicks for the warmth and protection this close huddle offers. The parents return daily with food, mainly fish and squid, and are able to locate their chick among hundreds of others. There is insufficient time before the onset of winter for the king penguin chicks to fledge and go their own way, so they remain with their parents in the colony for their first winter. For the winter months they huddle together for warmth, aided by their thick layer of blubber and a coat of brown down feathers, which made the early Antarctic explorers think they were another species of penguin. The parents feed them only occasionally and they must rely mainly on stored food all winter. The adults return in spring and then feed their chicks frequently, and they are finally fledged and ready for life at sea by midsummer of the following year, having been dependent upon their parents for eleven months. This lengthy raising period means that king penguins are only able to breed in alternate years. Both parents moult after their chick has left and before they prepare to nest again. (See the color insert.)

Gentoo Penguin (*Pygoscelis papua*)

A medium-sized penguin measuring 30 inches (76 cm) and weighing 12 pounds (5.5 kg), the gentoo has typical penguin bluish-black upperparts and white underparts, but it is easily recognized by the irregular white patch over each eye, which is linked over the crown by a narrow white bar. Its bill is reddish-orange with black on the front of the top mandible, and its legs and feet yellowish-orange. The sexes are alike in plumage, but the male is slightly larger. The gentoo penguin has a circumpolar distribution, breeding on the tip of the Antarctic Peninsula, and on many sub-Antarctic islands, mainly those near the Antarctic Convergence, including Heard, Kerguelen, Crozet, Marion, Prince Edward, South Orkney, South Shetland, South Georgia, Falklands, Staten, and MacQuarie. Birds of the more southerly populations, such as those on the South Shetland and South Orkney Islands, migrate north for the winter months. Two subspecies are recognized: *P. p. ellsworthi* breeds on the South Sandwich, South Orkney, and South Shetland Islands, and on the Antarctic Peninsula; *P. p. papua* breeds on the other islands. The Falkland Islands have long been a gentoo penguin stronghold, but the increased settlement and development of the islands has disturbed the birds and they have lost much of their preferred tussock grass nesting areas to grazing animals. They now nest mainly on rocky shores, making a nest of stones, sticks, and a little grass collected from farther inland.

Gentoo penguins are sexually mature when only two years old, and return to their rookeries from August to October. They nest in small colonies, rarely more than several hundred pairs, and aggressively build and defend their nests, each

Gentoo Penguins   *One of the five species of penguins that live in the Maritime Antarctic, a milder region than the Antarctic Plateau, which comprises a small peninsula jutting north toward the tip of South America and many islands such as the South Orkneys and South Shetlands.*
Photo: Tony Hathaway, Dreamstime.com

stealing from their neighbors' nests while they are absent. They lay two eggs, which are incubated by both parents for the thirty-five-day incubation period, and although the second egg is laid four days after the first, they both hatch on the same day; and both chicks normally survive when food is plentiful. When they are five weeks old the chicks form a creche while the parents provide food, and at the age of thirteen weeks they are fledged and waterproof and can go to sea. When their chicks have left, the adults moult and depart soon afterward, but may first undergo a false nesting activity during which they rebuild the nest they used previously, or even make a totally new nest, before they head off to spend the winter at sea; this unusual behavior is believed to strengthen the pair bond. Some individuals winter as far north as the Peninsula Valdes on the coast of central Argentina, and vagrants have been seen on the coasts of Tasmania and New Zealand.

The gentoo penguin's diet is mostly krill, with small fish, crustaceans, and squid. They forage close to the shore, and rarely dive deeper than 300 feet (91 m). They are killed by leopard seals, sea lions, and killer whales while at sea, and on land their eggs and small chicks are taken by caracaras and skuas. Their total population is currently estimated at about 300,000 breeding pairs.

Macaroni Penguin (*Eudyptes chrysolophus*)

Named for eighteenth-century English dandies, the macaroni penguin is the largest of the *Eudyptes* penguins. It has bluish-black upperparts and head, white upper tail coverts and white underparts, and a black chin and throat that ends on the upper breast in a V shape. It has a reddish-brown bill, and its legs and feet are pink. The most distinctive feature of this species is the golden head plumes, which are joined on the forehead and extend from there along the sides of the crown and hang down behind the eye. When excited it raises its crown feathers and bristles the yellow tufts. It has a large, reddish-brown bill with a bare pink patch at its base. The similar royal penguin has a larger bill and a whitish-gray throat, instead of black, but has been considered by some ornithologists to be a race of the macaroni penguin, hence *E. chrysolophus schlegeli*. An adult macaroni penguin weighs 12 pounds (5.5 kg) and is about 27 inches (69 cm) long. Although the sexes cannot be identified by their plumage, males are sometimes larger and have heavier bills. The macaroni penguin breeds outside the Antarctic Circle, on the tip of the Antarctic Peninsula and its offshore islands, plus the islands of Kerguelen, Heard, Crozet, Prince Edward, Bouvet, South Sandwich, South Georgia, South Shetland, South Orkney, and the Falklands. Like the gentoo penguin, its southern populations migrate north to warmer latitudes for the winter, and wanderers have reached South Africa.

Female macaroni penguins are sexually mature when five years old, the males at six years. They breed in huge close-packed colonies on the beach, with a simple nest—just a scrape or shallow depression in the gravel or soil, or in tussock grass—every square yard. They return to the colonies between September and November, laying their eggs in the month after their arrival. Both parents share the duties of incubating the two eggs, which takes thirty-five days, one going to feed while the other keeps the eggs warm, but the first egg is small and often does not hatch. The male then broods the chick for its first three weeks while the female gets food for them both. Then the down-covered chick joins a creche of other youngsters that huddle together for warmth, and continues to be fed by both parents until it fledges at ten weeks and is ready to go to sea. When their chicks have left, the adult penguins return to the sea for several weeks to feed and regain their condition and then go back to their colonies to moult, departing again in April and May to spend the Austral winter at sea and returning only to the breeding site the following spring.

Macaroni penguins waddle rather than walk, holding their flippers out sideways and backwards while upright, and lowering them when they settle onto the nest. They forage close to the beach and rarely dive as deep as other penguins—usually no more than 150 feet (45 m)—but are still fair game for their main predator the leopard seal, and they hurl themselves 6 feet (1.8 m) up from the water onto the ice when the predators are in the vicinity. The macaroni penguin is the most abundant penguin in Antarctic and sub-Antarctic waters with an estimated 10–12 million breeding pairs, with 5 million pairs nesting on South Georgia.

Chinstrap Penguin (*Pygoscelis antarctica*)

This penguin has a bluish-black head with white cheeks, chin, and throat, which is crossed by a narrow black "chinstrap," and both male and female have the same plumage. It has a black bill, its legs and feet are pink, and it measures 28 inches (71 cm) and weighs 10 pounds (4.5 kg). There are no recognized subspecies. It is a bird of the northern Antarctic, occurring only within the Antarctic Circle on the Balleny and Peter I Islands.

The chinstrap penguin has increased in recent years, probably because of the greater availability of krill since the decline of the baleen or whale-bone whales. It is second to the macaroni as the most abundant penguin in Antarctic and sub-Antarctic waters, with an estimated population of about 7.5 million breeding pairs. Its largest rookeries are on the shores of the Antarctic Peninsula and on islands in the Scotia Sea—South Georgia, South Sandwich, and South Shetland. The largest colony, on Zavodovski Island in the South Sandwich group, consists of 10 million birds, and there are smaller rookeries on Heard, Kerguelen, and Bouvet Islands. It has wandered as far as Tasmania and MacQuarie Island, and is probably the most nocturnal of the penguins, as it feeds at night as well as during the day, but it does not venture far out to sea.

The chinstrap penguin returns to its nesting colonies in November, and the juveniles are ready to depart the rookeries in March. It has a faster breeding rate than the crested penguins of the genus *Eudyptes*, which show preferential treatment to the strongest chick and never raise more than one. Like them, the chinstrap lays two eggs in late November, but hatches and raises both chicks, showing no favoritism and feeding each one equally. Both parents share the thirty-five-day incubation duties—in seven- or eight-day shifts—and the chicks spend thirty days in the nest before joining a creche. They fledge and go to sea when they are sixty days

Chinstrap Penguin *A penguin of the northern Antarctic Ocean, which has increased to an estimated population of 7.5 million breeding pairs, most likely due to the great increase in krill since the demise of the balleen or whale-bone whales. It is the most plentiful species after the macaroni penguin.*
Photo: Tony Hathaway, Dreamstime.com

old. The breeding results are poor, however, when sea ice persists close to the nest colony and the parents cannot get to the open sea to feed. Starting in March they leave the breeding colonies and move north of the pack ice for the winter. The main predators of this species are leopard seals at sea and skuas on land.

Royal Penguin (*Eudyptes schlegeli*)

The royal penguin is a bluish-black bird with whitish underparts and a grayish-white face, chin, and throat. Its orange-yellow head plumes arise from the center and sides of the forehead and hang down the back of its head. It differs from the other species of crested penguin in having white or grayish cheeks and throat.

Australian Cassowary   This is now a rare bird of the rain forest of northeastern Queensland, but other races of the species are more plentiful in New Guinea and several of its neighboring islands. Mainly a fruit-eater, it is an important dispersal agent of the seeds of forest trees.

Photo: Courtesy Ralf Schmode

Takahe *The largest rail, the takahe was believed extinct for many years until rediscovered in New Zealand's Fiordland mountains in 1948. As its small population continued to dwindle there it became the subject of intense captive-breeding efforts by the Department of Conservation. It currently has a population of about 200.*
Photo: Clive Roots

Kagu *A unique and very rare flightless bird, with a total population of about 1,000, the kagu is endemic to the South Pacific island of New Caledonia. It still has large wings on which it glides down the forested mountain slopes, but its flight muscles have degenerated through lack of use.*
Photo: Courtesy Kenneth W. Fink

King Penguins *Three king penguins on a South Georgia, Antarctica beach. The second largest of the seventeen species of penguins, they are wide-ranging birds of the cold sub-Antarctic seas. Their current population is believed to be about 1 million.*
Photo: Tony Hathaway, Dreamstime.com

Emperor Penguins *The largest penguin, the emperor penguin is one of the three species that breeds within the Antarctic Circle, but the only one to do this in midwinter. As the chicks mature they congregate in creches under the care of a few adults, while the others go out to sea to fish.*
Photo: Courtesy NOAA

Rockhopper Penguin
Photo: Lynsey Allen, Shutterstock.com

Yellow-eyed Penguin
Photo: Clive Roots

Erect-crested Penguin
Photo: Courtesy Dave Houston

Fiordland Crested Penguin
Photo: Courtesy Dave Houston

Kakapo *One of the flightless birds that still has large wings but cannot fly due to the degeneration of its breast muscles, the kakapo climbs trees in typical parrot fashion—using its bill and feet—and clumsily "glides" back to the ground.*

Photo: Courtesy Department of Conservation, New Zealand. Crown Copyright. Photographer: Rod Morris, 1979

Snares Islands Penguins   *A small penguin restricted to the Snares, a group of sub-Antarctic islands south of New Zealand. Numbering about 30,000 breeding pairs, the species is considered quite secure as its island home is one of the few in the world that is free of introduced predators.*

Photo: Courtesy Dave Houston

Banded Rail   *The greatest natural colonizer of all birds, this rail is a native of Southeast Asia, Indonesia, and Oceania. It has reached many islands and is the ancestor of such flightless species as the now extinct Dieffenbach's rail and MacQuarie Island rail, and possibly the weka, plus several semiflightless subspecies like the rails of New Caledonia and the Auckland Islands.*

Photo: Sword Serenity, Shutterstock.com

Rifleman *A semiflightless species of native wren, and New Zealand's smallest endemic bird, the rifleman is arboreal but has very feeble flight, and can only flutter short distances. It "runs" up tree trunks searching for insects and then floats back down to the ground.*
Photo: Courtesy Department of Conservation, New Zealand. Crown Copyright. Photographer: M. F. Soper

Atlantic Puffins *These puffins of the northern Atlantic Ocean nest in burrows on the cliff tops. After the breeding season they fly out to sea and moult all their wing feathers, and although they are flightless for several weeks they can still swim fast enough underwater to catch fish, even with their reduced wing area.*
Photo: Tony Hathaway, Dreamstime.com

Walden's Hornbill   A male Walden's hornbill of the Philippines brings food to his partner, who is "imprisoned" in the nest cavity for the duration of egg laying, incubation, and chick raising as protection from predators. During this time she moults all her primary feathers at once.
Photo: Courtesy Jon Hornbuckle

An adult royal penguin weighs 10 pounds (4.5 kg) and is 29 inches (74 cm) long. It is one of the group of crested penguins, and is very similar to the macaroni penguin, with which it has been linked in the past as a subspecies, but is now generally accepted as a full species. It is slightly larger than the macaroni penguin, which also differs in having a black chin. As with all the crested penguins the sexes appear similar, with the males being slightly larger.

The royal penguin is one of three species endemic to just a single small island or group of islands; the Snares Island penguin and the Galapagos penguin being the others. The royal penguin nests in four colonies only on MacQuarie Island, south of New Zealand, where its population numbers about 2 million birds. The island has been a wildlife sanctuary since 1933, but both feral cats and rabbits live there also. The royal penguin does not breed until it is six years old, but apparently does not enjoy much success in raising young until it is ten years old. Each year the breeding penguins return to the same nest sites, and they usually have the same partner throughout their reproductive lives. Males arrive at the breeding islands first in September and are followed soon afterward by the females. Their nesting behavior is typical of the crested penguins in which two eggs are laid, but the first is smaller and may not hatch; if it does the chick is small and weak and does not survive. Both parents share the incubation period of thirty days, and the males guard the chicks initially while the females bring food; when they are four weeks old all the chicks join a creche. Most adults, having been relieved of the daily guard duty, are then able to fish and feed the chicks, the diet being mostly euphausids. The chicks fledge at the end of February, leaving the parents free to feed just themselves and regain their lost weight in readiness for their moult in March, after which they return to the sea until the next breeding season. Royal penguins are migratory birds that leave MacQuarie Island at the end of the breeding season and have been sighted throughout the southern oceans, from Tasmania to the Antarctic regions.

Temperate Climate Penguins

The subtropical convergence is the limit of the cool sub-Antarctic water flowing north. It lies about 1,000 miles (1,600 km) north of the Antarctic Convergence and roughly parallels it except for bulging out to bathe several oceanic islands, New Zealand's South Island, and the southern coasts of South America with cool water. A number of major island groups lie in these seas. They include the Falklands, Tristan, Amsterdam, and St. Paul, and the many islands south of New Zealand including the Auckland, Chatham, and Campbell Islands. Technically, these islands are considered sub-Antarctic as they are all situated south of the subtropical convergence, but despite being cooled by the northerly flowing water, they enjoy a far more temperate climate than the islands farther south. Therefore, for the purposes of this book, the penguins that nest on these islands are considered temperate climate species. After all, if water temperature was the sole criterion for categorizing the distribution of penguins, even the white-flippered penguin, which occurs only on the coast near Christchurch, New Zealand, must be considered a sub-Antarctic species; while neither the Humboldt penguin of the Peruvian coast nor the Galapagos penguin—which actually lives on the equator—could be considered tropical birds.

Six species of penguins inhabit the cool oceans between the two convergences. Four of these, the erect-crested, Fiordland crested, Snares Islands, and yellow-eyed penguins live on New Zealand's South Island or its neighboring offshore islands. The Magellanic penguin has a much wider range, including the temperate islands of the Falklands and Tierra del Fuego and both the Atlantic and Pacific coasts of southern South America. The remaining one, the crested rockhopper penguin, has probably the widest range of all the penguins, occurring in all the oceans in the vicinity of the Antarctic Convergence, and ranging north to Tristan da Cunha at 38°S in the Atlantic Ocean and to the St. Paul and Amsterdam Islands at the same latitude in the Indian Ocean.

Erect-crested Penguin (*Eudyptes sclateri*)

This is the penguin with the fully erect, "brush-cut" golden-yellow crests. The parallel crests begin at the gape of the bill and extend over the eye to spike outward at the back of the head. It has the usual blackish upperparts and whitish belly, and its head is mainly black, while the bill is reddish and the legs and feet are pink. The erect-crested penguin is 25 inches (63 cm) long and weighs just over 8 pounds (3.6 kg) when adult, and the sexes are similar externally, although males may occasionally be larger and have heavier bills. There are no subspecies. It is similar to the Fiordland crested penguin and the Snares Islands penguin, differing only in having the complete yellow crest tufting upward rather than flat on the nape. Also, its crest begins at the base of the upper bill rather than at the gape. The erect-crested penguin also has a narrow strip of bare skin along the edge of the bottom bill.

This species breeds on several groups of sub-Antarctic islands south of New Zealand, with large colonies on Bounty Island and the Antipodes Islands, and smaller ones on the Campbell and Auckland Islands. Throughout its range it associates with the rockhopper penguin except on Bounty Island. Its dispersal range during the winter months is uncertain, but it has been seen from Cook Strait to MacQuarie Island, and has also reached the coasts of Tasmania and South Australia. The erect-crested penguin is a colony breeder, nesting on the island's rocky coastlines and also on ledges accessed after a steep climb up the cliff face. The breeding season begins in September when the males return to the islands and stake their claim for a nest site, usually in the face of stiff competition, and make simple nests of stones and soil. The females then arrive and lay two eggs, the second coming several days after the first and almost 50 percent larger, which is characteristic of the genus *Eudyptes*. Like the other species in the genus they practice "brood reduction," but with a difference. In the others both eggs generally hatch but the weakest chick dies, whereas in the erect-crested penguin only the second-laid egg hatches, and the first, smaller one is pushed to the side of the nest or even outside it. The chick is fledged by February when it immediately takes to the sea to feed, and is followed by the adults in April and May after they have moulted. They swim far out to sea to feed but are believed to be surface feeders of small fish and krill rather than deep divers like most of the penguins. The erect-crested penguin's numbers are not known but it appears to be decreasing for unknown reasons, despite the lack of introduced predators on the nesting islands. (See the color insert.)

Fiordland Crested or Thick-billed Penguin (*Eudyptes pachyrynchus*)

A smaller species, weighing 7 pounds (3 kg) and measuring only 23 inches (58 cm) long, the Fiordland crested penguin has a bright-yellow crest on each side of its head, beginning at the base of the bill and sweeping over the eye to the back of the head, but lying flat on the nape, unlike the raised crest of the erect-crested penguin. A more distinguishing feature is the parallel white stripes on its black cheeks, numbering up to five on each side. The sexes are alike in plumage, although the males may be larger and have heavier bills and more distinct cheek stripes. No subspecies are recognized. This penguin has a limited nesting area, occurring only in New Zealand's south Westland, Fiordland, on the west coast of Stewart Island, and on the islands in between such as Solander and Whenua Hou Islands. It is now one of the rarest species, with an estimated total population of only 3,000 pairs.

The Fiordland crested penguin nests individually or in small colonies, usually in caves, in dense undergrowth under tree roots, and among wet moss-covered boulders along creek beds, often some distance from the shore, possibly to escape the hordes of sandflies for which its habitat is notorious. The region receives very heavy rainfall, about 15 feet (4.5 m) annually, and nest sites have been described as muddy bogs. After spending the winter at sea the penguins come ashore in July and breed the following month. Two eggs are laid, first a small one, then three or four days later a much larger one. Both parents share the incubation, which takes six weeks, but although both eggs may hatch, the chick from the first egg is weak and unable to compete with the larger chick from the second egg, and does not survive. At first the chick is guarded by the male while the female provides the food, but both soon need to fish to keep up with the demand. After receiving the sole attention of its parents this chick is fledged and ready to depart to sea in November when it is about ten weeks old, but it will not be sexually mature until five years of age when it returns to the home colony in search of a mate. Adult penguins stay in coastal waters until January, when they return to their nesting sites to moult, finally moving out into the open sea in March. Their exact winter range is unknown, but they are most often seen in the seas off southern Australia and Tasmania. This penguin's diet comprises krill, squid, octopus, and small fish.

In the late nineteenth century pioneer conservationist Richard Henry of Resolution Island reported thousands of Fiordland crested penguins crowding the small caves in the rocks above the shoreline. Since then, like all the penguins that nest on New Zealand's main islands, the species has suffered heavily from predation by dogs, ferrets, and cats. (See the color insert.)

Snares Islands Penguin (*Eudyptes robustus*)

A small penguin, weighing just over 6 pounds (2.7 kg) and measuring 21 inches (53 cm), the Snares Islands penguin is similar to the Fiordland crested penguin but is darker in color, has a heavier bill with bare skin at its base, and lacks the white cheek markings. The sexes are alike in plumage, although the males on average are larger and have even heavier bills. There are no subspecies. This penguin breeds only on the Snares, a group of sub-Antarctic islands with a land area of only 840 acres (340 ha) lying southeast of New Zealand, islands on which the dominant vegetation is large tree daisies of the genera *Olearia* and *Brachyglottis*. These penguins are colony nesters, their colonies usually containing up to 1,500 pairs, in open areas in the "daisy forest" and scrub and on bare rocks along the shore. The males return to the islands from the sea in August and immediately set about building the nest of stones, sticks, and earth. The female then arrives and lays two eggs, and like the other crested penguins the first egg is smaller than the second egg and only one of the hatchlings survives. Neither of the adults have fed during this period of nest building and egg laying, so after about ten days of shared incubation the male returns to the sea for almost two weeks to feed, then returns to relieve the female. When the chick hatches it is brooded by the male for three weeks, while the female goes to sea daily to acquire food for it. The Snares Islands penguin eats euphausids (small crustaceans), squid, and small fish. When it is three weeks old the chick joins a creche while both parents undertake the daily fishing routine to provide sufficient food, locating their chick among dozens with ease. The chicks have all their feathers in January when they are about eleven weeks old and then depart for their independent life at sea; they do not return to the colony to find mates until they are sexually mature at the age of five or six years. When their chick has gone, the parents are able to regain their condition and then they moult, returning to the sea in April. Their winter range at sea is not exactly known, but birds have been seen as far north as Australia.

The Snares Islands are completely free of introduced terrestrial predators so the penguins are considered as safe as a species restricted to a single group of islands can be, but the islands have been off-limits to visitors since 1961 to prevent the accidental introduction of pests. The penguin's only predators are therefore natural ones—the leopard seal, sea lion, and the skua. Their current population is in the region of 30,000 breeding pairs. (See the color insert.)

Yellow-eyed Penguin (*Megadyptes antipodes*)

The yellow-eyed penguin is a rare New Zealand species that breeds on the southeast coast of South Island and on the sub-Antarctic Auckland and Campbell Islands. It is a medium-sized penguin, weighing 12 pounds (5.4 kg) and measuring 26 inches (66 cm), with typical penguin plumage of slate-gray upperparts and a whitish chest and belly. However, true to its name it has yellow eyes and its forehead and crown are pale golden-yellow streaked with black, and its face, cheeks, and throat are similar although more brownish. A band of pale-yellow feathers beginning above its yellow eye circles the hind-crown. The bill is dull red and the legs and feet are pink. Young birds lack the yellow stripe until they attain full plumage at one year of age. The sexes are alike externally and there are no subspecies. The yellow-eyed penguin nests only on South Island's east coast, from southeast Otago north to Cape Wanbrow, and on Stewart, Codfish, Snares, Auckland, and Campbell Islands. The destruction of the coastal forest, which provided nesting sites, and the loss of its chicks to stoats and feral cats have contributed to its rapid decline, and it is now very rare, with a population estimated at about 2,000 pairs.

The yellow-eyed penguin has also suffered from unaccountable "die-offs," such as the one that occurred during the summer of 1989–1990 when one-third of the 400 adult penguins on the Otago Peninsula died from unknown causes, and the 120 orphaned chicks were hand-raised. The Yellow-eyed Penguin Trust is trapping predators, has fenced their nesting site, and is conducting an education program warning people to prevent their dogs from attacking penguins, and not to dump unwanted cats in the area. Another reason for the drastic decline of the species was the use of set nets in the coastal waters where it feeds. Fifty penguins are known to have been caught and drowned in recent years, and to assist the birds' recovery the use of these nets off the southern coast of the Moeraki Peninsula in Otago was banned in 1991.

Unlike the Fiordland crested penguin, the yellow-eyed penguin is a sedentary bird, staying in the seas adjoining its nest sites and only rarely straying north as far as Cook Strait. It is also less colonial than most penguins, often nesting singly or in loose colonies away from the beach in forest and scrub, widely spaced, out of sight of their neighbors. The nest is often a platform of sticks and grass, and may be in thick bush under tree roots or logs. Returning to its nesting areas in August, the yellow-eyed penguin breeds in September and October, laying two eggs that are incubated by both parents for forty-five days. Chicks are brooded for the first two weeks and fed by regurgitation by the nonbrooding parents, who come ashore at night and follow well-beaten tracks to the inland nests. The young leave the nest when about four weeks old, and are guarded by a parent for a further three weeks, after which they are left on their own as their demands for food require both parents to go fishing. They are fledged and begin to leave the nest sites in February. They are deep divers when searching for their food, which is mainly fish and squid, often swimming a long distance from shore to do this, and staying submerged for up to three minutes. On the mainland the loss of coastal forest breeding habitat and predation by cats, weasels, stoats, and dogs are the major reasons for this species decline. On Auckland Island they are now safe from predators; both the rats and feral pigs there have been eradicated by the Department of Conservation. (See the color insert.)

Rockhopper Penguin (*Eudyptes chrysocome*)

The rockhopper penguin weighs up to 7 pounds (3.1 kg) and measures 22 inches (56 cm) long. It has white underparts, its head and face are blackish-gray, darker on the crown, and the rest of the upperparts are bluish-black. Its occipital crest—a narrow line of pale-yellow tasselled plumes hanging down each side of the head—does not meet over the forehead but extends individually from behind the nostril, above the eye and along the sides of the crown. Its bill is reddish-orange and its legs and feet are pink. It also has bright-red eyes, whereas in the other species they are a dull dark red. The sexes are similar although the males may be larger and have heavier bills.

The rockhopper penguin has a circumpolar distribution, in which three subspecies are recognized. The subspecies E. c. chrysocome breeds on Tierra del Fuego, islands off Cape Horn, and the Falkland Islands. The wider ranging race, E. c. filholi, breeds on the islands of MacQuarie, Campbell, Auckland, Antipodes, Snares, Bounty, Kerguelen, Marion, Heard, Crozet, and Prince Edward. The breeding range of the slightly larger E. c. moseleyi includes Tristan da Cunha, Gough, St. Paul, and the Amsterdam Islands. Although the ranges of the rockhopper penguin and the macaroni penguin overlap, the rockhopper also occurs on more northerly islands beyond the range of the macaroni penguin, which extends farther south to the Antarctic Peninsula. The winter pelagic range is not exactly known, but the rockhopper certainly ranges widely and is a fairly common visitor to the coasts of Australia from Freemantle to mid–New South Wales, and is occasionally seen on the South African coast.

Male rockhopper penguins of the northern populations, on MacQuarie and Gough Islands, return to the islands to claim a territory in September, and build a nest of stones and vegetation. The females arrive soon afterward and lay two eggs, and the young join a creche at the age of four weeks and are fully fledged and ready to begin their life at sea in March. The more southerly colonies are generally about six weeks behind this schedule. The rockhopper is still a common species, but its numbers are decreasing for unknown reasons, despite the absence of alien predators on their islands. The total population is currently about 3 million birds, with two-thirds of these nesting in the Falkland Islands.

Magellanic Penguin (*Spheniscus magellanicus*)

This penguin was named after Ferdinand Magellan, who was the first European to see it when he rounded Cape Horn in 1519. It is one of four species in the genus *Spheniscus*, which are all very similar, noncrested black-and-white birds with wedge-tails, that are further characterized by their loud, donkey-like braying. A medium-sized bird, weighing 8 pounds (3.6 kg) and measuring 27 inches (69 cm) in length, it has slate-gray upperparts and a black head with a broad white band from the base of the bill, over the eye and joining at the throat. There is a wide black band across the white upper breast, and another, thinner, black band crosses the breast below it and extends down the flanks to the thighs in the shape of a horseshoe. It has a heavy black bill crossed by a gray bar and its feet and legs are black. In size and plumage the sexes are alike and there are no subspecies. In the summer breeding season they moult the feathers around their eyes and the exposed skin then turns pink. They regrow these feathers as winter approaches.

The Magellanic penguin has the most southerly distribution of the Spheniscid penguins, nesting on the Patagonian coast of Argentina from Peninsula Valdes (42°S latitude) south to Cape Horn and then north along the western coast and islands of Chile to Concepcion. It also breeds on the Falkland Islands and Juan Fernandez (Robinson Crusoe) Island. Some birds migrate northward to escape the southern winters, reaching southern Brazil on the east coast and central Chile on the west, where they overlap the range of the Humboldt penguin. Since the decimation of the latter species, and the great reduction of the black-footed penguin, the Magellanic penguin is the most plentiful of the Spheniscids. Its largest colonies are on the Argentine coast, the one at Puerto Tomba containing over 1 million birds.

Magellanic Penguins *A Magellanic penguin and its chick at the entrance to its nest hole. A temperate-climate species, it nests mainly on the coast of Argentina and on the Falkland Islands. It is the most plentiful of the Spheniscid penguins, the famous and important colony at Puerto Tomba containing 1 million birds.*
Photo: Alexander Mikula, Dreamstime.com

The Magellanic penguin has a much shorter breeding season than the other members of its genus. It returns to its nest sites in August and burrows into the sand for several yards, enlarging a nest cavity at the end of the tunnel which it lines with grass and sticks. Two eggs are laid and both parents share their incubation for forty days, and then take turns to either guard the chicks or go fishing for them. When they take short-cuts back to the sea they dive off headlands sometimes 20 feet (6 m) high, bouncing off the rocks before reaching the water. Both chicks are usually raised, barring the heavy rainstorms that often flood the burrows and drown the young, but they may nest again if there is time left in the season. The chicks are feathered and independent in March and April at the age of sixty days.

Although oil spills and overfishing have affected this penguin, and it is also held in check by sea lions, killer whales, gulls, and caracaras, it is not considered to be in any serious danger at present.

Warm Latitude Penguins

For birds that require both cool water and low land temperatures, a range that extends north to the equator may seem anomalous, but penguins live only where waters originating from Antarctica provide the correct temperature and a rich source of food, however high the latitude. With access to cold water they can withstand high land temperatures because of their evolved adaptations, and they live in tropical or subtropical latitudes only because of the coolness of the water. Warm equatorial waters above 68°F (20°C) are a barrier to their northward progression. The most northerly point of penguin distribution—the Galapagos Islands—despite straddling the equator have a temperate climate because their waters are cooled by the Peruvian Current. Flowing north from Antarctica along South America's west coast the current is forced away by the northwest continental bulge of Ecuador and reaches the Galapagos Islands, 465 miles (750 km) away in the Pacific Ocean. Two of the four species of Spheniscid penguins occur in this region. The Humboldt penguin lives on the coast of Peru to the point where the current swirls away into the open ocean, and the descendants of those penguins, which followed the current years ago and safely reached the islands, evolved into the Galapagos penguin. A third Spheniscid, the black-footed penguin, is also a warm latitude bird. Living on South Africa's west coast, it ranges north well into the tropics to the coast of southern Angola, following the north-flowing Benguela Current, which is an offshoot of the cold west wind drift circling the southern oceans. On South Africa's east coast the penguin does not extend as far north because the major water flow there is southerly and warm, arising from the equatorial region.

On Australia's west coast the blue penguin's range extends to the Tropic of Capricorn as the West Australian Current carries cool water there from the Antarctic. As in southern Africa and South America, the natural flow of the northern offshoots of the West Wind Drift is along the western coasts of the southern continents. Except for the tip of South America the eastern coasts of the southern continents are warmed by south-flowing currents emanating from the equator, and soon become too warm for penguin habitation. The cold Falkland Current, which flows north along the east coast of South America, meets the south-flowing warm Brazil Current off the mouth of the Rio de la Plata, allowing the Magellan penguin to colonize this coast as far north as Peninsula Valdes, so it is considered a temperate species.

Black-footed, Jackass, or African Penguin
(*Spheniscus demersus*)

The black-footed penguin is the only African penguin, a small bird measuring 26 inches (66 cm) long and weighing 7 pounds (3.1 kg) when adult. It is blackish-gray above and white below, with a narrow black inverted horseshoe band crossing the upper breast and extending down to the thigh. Occasionally a second partial or even complete black band crosses the white foreneck, similar to the markings of the Magellanic penguin. The head is black with a white band extending around the sides of the crown from the base of the bill over the eye and looping around behind the cheeks to meet the white breast. The bill is thick and black, with a transverse gray bar, and the legs and feet are black. The sexes are alike externally and there are no subspecies.

The black-footed penguin lives on the coasts and islands of South Africa from the Cunene River (the border between Angola and Namibia) around the Cape of Good Hope to Maputo on the coast of Mozambique, at the same latitude as Pretoria. It breeds on two dozen islands off the coast of southern Africa, including Dassen, Marcus, Malgas, Bird, St. Croix, Dyer, and Robben. It has recently begun to colonize the mainland, at Betty's Bay, Simonstown, and on the Namibian coast. It is unclear why it has chosen to breed on the mainland, where it is vulnerable to predation from jackals, hyaenas, vultures, baboons, and even leopards, but it is certainly safer from oil spills, and fur seals do not commandeer the beach space.

Like the other members of its genus, the black-footed penguin nests in large colonies, sometimes in the open but generally in burrows in the sand or in guano or beneath rocks and vegetation, lining its nest with feathers and seaweed. It breeds throughout the year, although mainly in the summer, laying two eggs that are incubated for thirty-nine days. Both parents take turns incubating and fishing for anchovies, sprats, and squid to raise their chicks. Compared to most species the black-footed penguin is a sedentary bird, fishing in the offshore waters in the vicinity of its nesting islands, and not given to wandering the open seas for months at a time. It was very common early in the twentieth century, with a population of almost 3 million, with the largest colony on Dassen Island containing 1 million birds. That colony now has only 30,000 birds and the total population is about 120,000 birds. The disturbance of their nest sites during the removal of guano, extensive egg collecting, and oil pollution have all contributed to their decline. The limited availability of breeding areas has also been cited as a possible reason, although these same areas at one time provided sites for many more breeding pairs; but there is now competition from the South African fur seal (*Arctocephalus pusillus*) on the island beaches.

The black-footed penguin has suffered more from marine oil spills than any other penguin, in fact, more than any other bird in proportion to its numbers. The stormy waters and islands off the Cape of Good Hope form a dangerous sea route, and many vessels have foundered. Major disasters in recent years include the wreck of the ore carrier *Apollo Sea* in June 1994 just 20 miles (32 km) north of Cape

Town, soon after it had refuelled with oil; 10,000 penguins were oiled but almost 50 percent were treated and recovered. Then in June 2000, the iron ore ship *Treasure* sank in Table Bay and leaked 1,300 tons of fuel, which formed a slick around Dassen and Robben Islands. Thousands of penguins and their chicks were threatened, and many were airlifted to Port Elizabeth, from where they were released to swim back in their own time; they completed the 620-mile (1,000 km) journey in three–four weeks, by which time the slick had moved away. Many oiled birds were also treated and chicks were taken from their nests to prevent their starvation. Fences were erected around the nest sites to prevent birds from going to sea to feed and getting oiled in the process. Once again, the South African National Foundation for the Conservation of Coastal Birds (SANCCOB) achieved a high rate of success in rescuing and rehabilitating oiled penguins.

Humboldt or Peruvian Penguin (*Spheniscus humboldti*)

The Humboldt penguin is a slightly shorter and stockier bird than the black-footed penguin, measuring 26 inches (66 cm) long and weighing 8 pounds (3.6 kg). Its head and throat are black, with a narrow white band extending from the base of the bill along each side of the crown and over the eye, becoming broader on the white breast. The upperparts are blackish-gray and the underparts white, with a narrow black band shaped like a horseshoe, beginning under the chin, crossing the chest, and running down each flank. Its markings are therefore similar to those of the black-footed penguin with one band across the upper breast, unlike the Magellanic penguin, which has two bands. It is shorter than the black-footed penguin, but has longer flippers. A sedentary species living in coastal waters near its rookeries, it breeds throughout the year, nesting in caves, crevices, among rocks, and burrowing into the sand (and guano when it was more plentiful). It ranges from islets near Valparaiso, Chile, where it overlaps the territory of the Magellanic penguin, to a few degrees south of the equator on the coast of northern Peru. This species and the black-footed penguin have always been the most popular and most easily obtained penguins for zoo exhibition, and it has been estimated that about 500 birds were exported annually from Peru up to 1977 when the practice was prohibited. In 1982 it was included in Appendix I of CITES, the international convention that controls trade in endangered species.

The Humboldt penguin breeds from March to December, with two broods in that period, followed by a two-month layoff to regain condition and undergo the annual moult. It is a very social bird that nests in colonies, laying two eggs about three days apart, and the parents share the forty-day incubation period and chick-brooding and feeding duties. Both chicks survive if there is plenty of food available, which is now rarely the case, and they fledge when three months old. Anchovies are their main source of food and they hunt close to the shore, rarely diving deeper than 200 feet (60 m). There are estimated to be only about 12,000 breeding pairs left, their rapid decline over the last few decades hastened by the El Niños, which affect their food supplies and produce heavy rainfall that floods their nests. They have also suffered from considerable human interference—overfishing for anchovies, the removal of guano, and entanglement in fishing nets—in addition to their natural predators, gulls, vultures, sea lions, and killer whales.

Blue, Little, or Fairy Penguin (*Eudyptula minor*)

The commonest of the Australasian penguins and the smallest species, the blue penguin weighs just over 2 pounds (900 g) and is 16 inches (40 cm) long. It has typical white penguin underparts, but its back and head are the palest of all species, being metallic slate-blue, which takes on a brownish hue as the bird ages. The sides of its face are gray, its chin and throat are white, the bill blackish-gray, and its feet and legs pinkish-white. The sexes are outwardly alike, although the males may be larger and have heavier bills. Three subspecies are recognized. *Eudyptula minor novaehollandiae* nests on the coasts of southern and southwestern Australia and Tasmania. *Eudyptula m. minor* breeds on the coasts of New Zealand's three main islands and also on Chatham Island. The white-flippered penguin (*Eudyptula m. albosignata*) (sometimes considered a separate species) breeds only on the Banks Peninsula near Christchurch, on Motunau Island, and on the coast of north Canterbury, with a current population of less than 4,000 pairs.

The blue penguin is a nonmigratory bird, fishing in the seas close to its nest sites and returning to them at nightfall. It is rarely seen due to its small size and

Humboldt or Peruvian Penguin   The Humboldt penguin lives on the coast of western *South America from central Chile to northern Peru—almost to the equator. Decimated by the loss of its food from overfishing, from the El Niños that flood its nests, and from overharvesting for zoos, only about 12,000 pairs now survive. Its bare facial areas help to dissipate heat when it is on land.*
Photo: Adrian T. Jones, Shutterstock.com

nocturnal habits, except where it has become a tourist spectacle when it returns to its burrow at night to relieve its mate or to feed the young. However, as they have shown at several sites, such as Tasmania's resort island of Bruny and the more famous "penguin parade" colony on Phillip Island south of Melbourne, they are not deterred by bright lights and spectators as they waddle up from the beach and cross roads to reach their burrows. The blue penguin has a long breeding season, laying its two eggs between July and January, generally in sand burrows, which may be 10 feet (3 m) long, while some birds nest close to the beach and others trek up to 1 mile (1.6 km) inland to their nests. It is also an opportunist and nests in caves, crevices, under dense vegetation, beneath buildings, and under wharves, even in Hobart's busy harbor. Nest boxes have been provided for them at many breeding sites.

This penguin's incubation period is thirty-six days and both parents share it and the feeding duties. Despite their swimming ability, turbulent seas take a great toll of blue penguins, especially young birds making their first attempt to survive in the ocean. They are often killed in large numbers by heavy seas and crashing surf, and many bodies are washed ashore during gales. They may travel 30 miles (50 km) to fish daily, but in winter stay at sea for several days and travel much longer distances. Their main natural predator is the brown skua, but they have suffered considerably from dogs and introduced ferrets and stoats.

Galapagos Penguin (*Spheniscus mendiculus*)

The Galapagos penguin is the fourth Spheniscid penguin, with a body length of 20 inches (51 cm) and weighing about 5 pounds (2.3 kg), making it the second-smallest species after the blue penguin. It is also one of the rarest penguins, with a population that rarely exceeds 800 pairs, and like all the native fauna of the Galapagos Islands it is fully protected by the government of Ecuador. It has blackish-gray upperparts and a black head, and a narrow white line extends from the eye, curving down behind the cheeks and joining on the black throat. A broad black band separates this white line from the whiter underparts, and beneath this band there is a narrower, indistinct black band that extends across the upper breast and down the flanks to the feet. The sexes are alike in plumage and size, and there are no subspecies. It is the most northerly penguin and the only species to breed on the equator; it is quite obviously descended from Humboldt penguins, which migrated from the Peruvian coast. A sedentary bird, it is restricted to the Galapagos Islands, where it breeds on the western sides of Fernandina and Isabela Islands, although immature birds can be seen off the other islands, especially Bartolome and Rabida. Adults seldom stray far from their nest sites, although stragglers have reached the coast of Ecuador.

The Galapagos Islands are cooled for much of the year by the cold waters of the Peruvian or Humboldt Current, whose rich upwellings against the islands brings nutrients that feed huge concentrations of anchovies and the larger fish that feed on them. But this upwelling is severely affected by the periodic El Niños, which warm the waters and dramatically reduce the fish. The penguins have been seriously affected by this phenomenon, while also contending with their natural predators such as the Galapagos hawks, which take penguin chicks, and the sea lions, which chase them in the water. In addition, alien rats are also capable of killing the penguin chicks. Fortunately, they have been able to rebound quickly after a bad year, such as the El Niño of 1982/83 when 70 percent of the population died due to the disruption of their food supplies by the warm water.

The penguins nest both colonially in small groups and in widely separated pairs, usually in a hole in the rocks or in a simple burrow. The two eggs are laid in typical penguin fashion, usually with a three–four-day interval between them, and they hatch at the same interval, which improves the survival chances of the first and therefore stronger chick when times are hard. The incubation and chick-raising duties are shared, and the chicks can be fed daily as food is available close to shore. Normally, only one chick is raised to maturity, which occurs very early—when they are only two months old.

Their behavior and adaptations allow the Galapagos penguins to live on the equator, for even if the air temperature may exceed 100°F (37.2°C) the water around the islands is cool, usually 60–70°F (16–21°C), and to reduce their body temperature they need only go for a swim. During the day on land they hold their flippers out to increase heat loss and to shade their feet. Small body size is favored in such an environment because of its greater effectiveness in dissipating heat, and it needs less to sustain it during times of reduced fish supplies. The larger size of the Antarctic penguins results from a rich and dependable food source despite the harsh climate, allowing them to reach triple the size. The Galapagos penguin has more extensive bare areas on its face to help lose heat, and a much thinner layer of body fat than the cold-climate penguins. Unlike those living in cold regions it has a variable breeding cycle that allows it to take advantage of a good year. Also unlike those species, the Galapagos penguin moults twice annually, possibly due to the wear and fading of its plumage by the tropical sun.

6 Divers and Dabblers

Penguins are not the only flightless waterbirds; several other aquatic species belonging to a number of bird orders can no longer fly. They lost the ability because they could find food and mates without flying, and because predation was either low or non-existent in their habitat; but unlike the penguins, they vary considerably in their appearance and habits. The grebes and cormorants are true divers, propelling themselves after fish with their lobed or webbed feet; and the giant coot, mainly a surface-feeding vegetarian, is also an adept underwater swimmer. The carnivorous flightless steamer ducks feed on the surface, but also dive capably to reach bottom organisms using both their feet and wings underwater; and two forms of flightless teal are surface-feeding or dabbling ducks that seldom dive. None of these waterbirds are as specialized as the penguins, which replaced their wings with paddle-like flippers. Their ancestors used their feet for swimming and they behave in much the same way in the water, but are now unable to fly. Until recently, these birds were equally represented, with five marine or partially marine forms and five that lived in freshwater lakes, but the latter have suffered more from human progress and two are now extinct.

■ MOUNTAIN LAKES SPECIES

For a coot and several grebes the right conditions for flightlessness were provided not by remote oceanic islands, but by isolated freshwater lakes high in the mountains of Central and South America, where food was plentiful and migratory flights unecessary. Their highly aquatic lifestyle provided relative safety from terrestrial predators, while their diving and swimming skills offered reasonable security from diurnal birds of prey. Their most serious predation probably came from carnivorous fishes large enough to take the chicks, but the males' habit of carrying their young on their backs reduced even that risk. These remote lakes were the land masses equivalent to the distant sea islands, and actually gave rise to half of the ten species of flightless continental birds, the other five being the large ratites.

The Species

Grebes

Grebes are duck-like birds with long, thin necks and pointed bills, and such rudimentary tails that they appear virtually tail-less. They are shy, almost totally aquatic birds that sink below the surface to escape close observation. Their oil glands are well developed and they have very dense waterproof plumage with up to 20,000 feathers each. Their breast pelts, known in the trade as "grebe fur," were once used for making muffs in Europe and saddle blankets in South America. Grebes catch most of their food underwater, yet seldom remain submerged for more than a minute. Diving from the surface or sinking with barely a ripple, they swim powerfully with their feet, which are situated at the rear of the body, an adaptation for diving and underwater swimming. They also have stiff flaps or lobes on their toes, which increases the surface area of the foot and aids swimming, and flexible tarsometatarsal joints between their legs and toes, which improves maneuverability. Grebes seldom leave the water and they nest on a bed of floating vegetation anchored to growing plants. They have the unusual habit of eating their own feathers and also feeding them to their chicks, possibly for their value as roughage.

Grebes have small wings and are weak fliers, but like the rails, several species make long, migratory flights twice annually. They are mostly freshwater birds, although in temperate zones they winter along the coasts. Even the flying grebes become completely flightless for up to four weeks during their annual moult, when they gather in large groups on inland waters or close to the seashore. They have also shown that under the right conditions they can lose their flight permanently and four of the known twenty-two species became flightless, although two of these are now extinct. Another species—the Alaotra or rusty grebe (*Tachybaptus rufolavatus*) of Lake Alaotra in northern Madagascar—is believed to be a very poor flier and may even be flightless, but its true disposition will likely never be known as it is on the verge of extinction.

The flightless grebes act in exactly the same manner as their flying ancestors. Diving from the surface they use their powerful legs to propel them in pursuit of fish or to reach the lake bottom in search of other aquatic life, which includes amphibians, insect larvae, leeches, and molluscs.

Short-winged Grebe (*Rollandia microptera*)

Also known as the Titicaca flightless grebe, this species is the only endemic bird of the mountain lakes and rivers of southern Peru and Bolivia, namely, Lakes Titicaca, Umayo, and Poopo, and rivers in the Titicaca Basin. Its major habitat, Lake Titicaca, lies in a large basin in the Andes. It is the world's highest navigable lake for large vessels, and the second-largest lake in South America. It lies 12,500 feet

(3,800 m) above sea level between Peru on the west and Bolivia to the east. It averages 300 feet (90 m) deep, with its greatest depth being 930 feet (283 m). The lake has a surface area of 3,300 square miles (5310 square km) and is almost divided by the narrow Strait of Tiquina. Lake Titicaca has many man-made floating islands called uros, woven from the reed beds of cattails (*Schoenplectus totora*), which are still plentiful around the shore and provide a safe haven for the grebe. Trout have been introduced and are competing with the native fish, and the lake is increasingly contaminated from mining activities and the organic wastes generated by the almost 1 million people who live in the basin and farm along its shores.

The short-winged grebe is a small bird, about 11 inches (28 cm) long, with a ragged crest, dark-brown back, chestnut nape, and white chin and foreneck. Its underparts are also white, speckled with chestnut; its bill is red and yellow; and its feet are bright yellow above and black below. The sexes are alike in size and plumage and there are no subspecies. It has the typical grebe form, with a long and narrow head and bill, a slender neck, and an elongated body. It is a good swimmer and diver, using its large feet, which are set well back on the body, and its broad lobed toes for propulsion. When alarmed it skitters fast across the water surface for some distance, flapping its stubby wings. In typical grebe fashion it makes a nest on floating vegetation in the reed beds. The presence of food year-round, large reed beds in the shallows for hiding and nesting, and the relative lack of predation encouraged the ancestors of the present-day grebes to abandon flight and eventually their flying muscles withered, their breasts lost the keeled shape, and their wings became shortened. Although it is now disturbed by tourists, fishermen, and reed-collectors, it does not appear to be in any immediate danger. Its numbers have been variously estimated at 2,000 to 5,000 birds, but it is listed as vulnerable by the World Conservation Union. It is considered relatively safe for as long as its protective reed beds remain, as they provide security from hunters, who fortunately for the grebe prefer to hunt the larger and more colonial giant coot.

Junin Grebe (*Podiceps taczanowskii*)

The Junin grebe is one of the world's rarest birds, with an estimated population of less than 300. It is endemic to Lake Junin, a large, shallow body of water at an elevation of 12,000 feet (3,660 m) in the central Peruvian Andes, 100 miles northeast of Lima. The lake covers 127,000 acres (53,000 ha) and is rarely more than 12 feet (3.6 m) deep. The grebe is a very pretty bird, with grayish-brown upperparts and lighter, almost white, feathering below. It is about 13 inches (33 cm) long, with a short crest and bright-red eyes. The sexes are alike externally and there are no subspecies. It is believed that its ancestors, which were silver grebes, were isolated on the lake during the latest period of world glaciation, which ended about 10,000 years ago. With nowhere to go, few predators, and plenty of food available, the grebe did not need to fly, and eventually became flightless. The strange thing is, however, that the silver grebe, which does fly, also still lives on Lake Junin. Four Junin grebes were relocated to Lake Chacacancha in 1984, but they did not survive and are believed to have been caught in fish nets.

The grebe of Lake Junin is a sociable bird and in more plentiful times lived in groups of up to 12 individuals, but less than 300 are now believed to survive, and one recent estimate placed the number at only 50 birds. Its steady decline in the last century from a population of several thousand has been attributed to a number of factors. Chemical pollution from a nearby copper mine may have caused breeding problems, as less than half the males of breeding age reproduced during a study made in 1977–1978. Water-level fluctuations, as the mine drew water from the lake, left nests high and dry, and introduced trout have replaced native fish, affecting the grebe's natural food supplies. Local mining activities have contaminated the lake with many mineral pollutants, especially lead and copper, and in dry seasons the hydroelectric dams lower its water level, which affects the surrounding marshland and concentrates the pollutants. The grebe's decline coincided with a dramatic drop in the fish and frog populations in the lake, but it is unclear whether poisoning or starvation is to blame for their demise.

The Junin grebe nests in colonies in tall reeds, well away from any possible predation by foxes and rats, and lays two white eggs on a bed of floating vegetation, between November and March. It moves to the center of the lake when it shrinks during the dry season. Like all grebes, the male carries the young on his back, while the female dives and provides food for them all. The young grebes then catch insects on the surface, but as they mature they dive for fish and crustaceans.

Coots

Coots are aquatic members of the rail family. They are expert swimmers and divers and have rounded lobes on their toes similar to the grebes, which increases the surface of their feet and improves their swimming ability. Like all rails capable of flight, they take to the air as a last resort, usually swimming between the reeds or diving underwater when threatened. When they finally fly their feet patter the surface for some distance before they become airborne, when they fly low and strong. Coots are mainly vegetarians and eat the roots, seeds, and shoots of water plants, but they do not reject animal food and are known to eat small fish, other birds' eggs, and many forms of invertebrates. In plumage all the coots are very similar, their blackish-gray feathering being relieved by a large, white or yellow frontal shield and red beak and legs. Only the giant coot is flightless.

Giant Coot (*Fulica gigantea*)

This is a large coot, and like the flightless grebe it also evolved in remote freshwater lakes in the New World. It once had a much wider distribution, but is now quite rare and localized, occurring in northern Chile, western Bolivia, and southern Peru, where it is generally seen at elevations of about 12,000 feet (3,660 m); however, it has been recorded at Cotacotani Lagoon at 16,000 feet (4,800 m) in Lauca National Park in northern Chile.

The giant coot is the size of a small turkey, about 24 inches (61 cm) long, with dark-gray body plumage and a black head and neck, and a bare yellow frontal shield over a white-and-yellow top bill and red bottom bill. It has huge red legs and four-lobed toes with long nails. Despite its size and name it is slightly smaller than the horned coot, which has a similar distribution, but still flies. The giant coot is a vegetarian although the chicks are insectivorous initially. They are very territorial birds and the males fight each other viciously in defense of their territory. They nest on large platforms of vegetation that may measure 8 feet by 8 feet (2.5 m by 2.5 m) and can support a person; and on small islands in the deepest water, so they are safe from terrestrial predators other than man. The eggs are laid in a depression in the center of the platform between August and December, and the nest is reused annually after being renovated.

The giant coot has always been considered a flightless bird even though a pair in the collection of the late Jean Delacour at Cleres, in France, "flew" over a high fence during a storm. However, it is not unusual even for pinioned birds to become airborne when they are wind-assisted.

■ INSULAR SPECIES

Like the many rails that found paradise on distant sea islands, a number of waterbirds also reached remote specks of land and found their environments ideal for colonization. The combination of agreeable climate, lack of predation, and ample food either on the island or in the surrounding waters annulled the need to fly, and the result was predictable. Several, both marine and freshwater species, became sedentary and eventually lost their flight. Unfortunately, the two Northern Hemisphere species—the great auk and the spectacled cormorant—were both exterminated in the last century. Of the remaining five living forms, one is tropical and the others occur on islands in cool, temperate southern seas.

The Species

Cormorants

Cormorants belong to the *Pelicaniformes*, the order of birds that includes the gannets, tropic birds, darters, and of course the pelicans. They all have webbing between their four toes, whereas ducks' feet are only webbed between three toes. Cormorants are mostly coastal marine birds, although some nest on lakes in the very center of Canada, Europe, and Asia during the short summers. Both the cormorants and the related shags are large birds with long bodies and necks, streamlined for catching fish underwater with their long hooked bills. Captive cormorants in the Orient have traditionally been trained to fish for their owners, but wear a collar to prevent them from swallowing their catch. Surprisingly, for birds that spend so much time in the water, the cormorants do not have waterproof feathers, and stand with their wings held out to dry after being submerged.

The cormorants are a contradictory group of birds, similar to the rails in their flying ability. They have short wings and many species seldom fly far, while others migrate almost 2,000 miles (3,220 km) between their winter and summer ranges. As a family they seem predisposed to flightlessness, and two species took the opportunity. They colonized isolated islands, far removed from each other, surrounded by food-rich seas and with relative freedom from predation. The largest and certainly the most spectacular of these birds, the flightless spectacled cormorant of Russia's Komandorski Islands, was exterminated in the middle of the nineteenth century. The other, the scruffy-looking Galapagos flightless cormorant, still survives rather precariously on islands off Ecuador.

Galapagos Flightless Cormorant
(*Nannopterum harrisi*)

Until it opens its wings the Galapagos flightless cormorant seems a typical cormorant, with its legs placed so far back on its body that it stands almost upright. But its wings have degenerated and measure only about 20 percent of its total body length and appear to have been rough-plucked; and it has already completely lost its breastbone keel and the pectoral muscles associated with flight. It is one of the

Galapagos Flightless Cormorant   *Like its flying relatives, the flightless cormorant holds its degenerated wings out to dry after a swim. It now survives on only two islands in the Galapagos Archipelago, and rarely ventures far from the shore. It has suffered from the periodic El Niños that warm the water and reduce its food supplies, but its numbers rebound quickly when it takes advantage of the intervening good years.*

Photo: Courtesy Dr. Robert H. Rothman, Department of Biological Sciences, Rochester Institute of Technology largest cormorants, with a length of 38 inches (97 cm), a blackish-brown bird with a few scattered white feathers on the sides of its face and neck; and with gray facial skin and black legs and feet. Although its wing feathers are of similar structure to those of the flying cormorants, the body feathers are much softer and thicker. However, in common with the flying species, the Galapagos flightless cormorant still perches on a rock with its tatty wings extended to dry out. Although cormorants have poorly functioning oil glands, their modified wing feathers allow air to escape and water to penetrate, reducing their buyouancy when swimming underwater, while air trapped in their dense body feathers prevents them from becoming waterlogged.

When swimming on the surface the cormorant's body lies low in the water, with just the head and neck exposed, but when seeking fish, octopus, and eels underwater it becomes a pursuit diver, holding its useless wings close to the body and propelling itself very fast with its legs and feet, which are heavier and more powerful than the other twenty-eight species of cormorant. Its ancestry is uncertain, but it is probably descended from either the guanay cormorant (*Phalacrocorax bougainvillii*) or the olivaceous cormorant (*Phalacrocorax olivaceus*), which occur on the adjoining mainland coast, although both are approximately 10 inches shorter in body length.

The Galapagos flightless cormorant is one of the world's rarest waterbirds, with a population of about 800 pairs. Endemic to the Galapagos Islands, it lives only on the coasts of Fernandina and northwestern Isabela Islands, where the cold Peruvian Current upwells. Fortunately, Fernandina is free of introduced predators, but Isabela has rats, cats, dogs, and pigs, all of which are potential threats to ground-nesting birds. In the water, even though they rarely venture more than 100 yards (91 m) from the shore, they are at risk from sea lions, which are also known to kill penguins. However, the cormorants face an even greater threat in the natural capriciousness of the climate, which causes yearly fluctuations in the breeding intensity and success of seabirds in the Galapagos Islands.

Alarms that the great increase in tourism to the islands may be disrupting the birds were apparently unfounded, and the circumstance having the greatest effect on the flightless cormorant's survival is the availability of food. Fortunately, the species is opportunistic and takes advantage of the good years, as it suffers severely during the bad ones. The interaction of ocean currents and water temperature fluctuations provides those good years and the occasional disastrous ones, due to the occurrence of the El Niño phenomenon. This happens periodically when warm water from the central Pacific Ocean flows eastward and raises the temperature of the normally cold seas off Peru and the Galapagos Islands, reducing the fish stocks and the animals that rely on them. The cormorant's population was halved by the El Niño of 1982/83, but soon recovered, and by 1992 again numbered about 800 pairs. The Galapagos cormorants nest in small groups on the shore or on rocky islands from where they can walk ashore. They are very sedentary birds that rarely move far from where they hatched. Their nest is a simple bed of seaweed, sea urchins, and sticks, and they lay two or three whitish eggs, but usually only one chick is raised. Both parents care for the chick until it is almost independent, when the female leaves the male to complete the raising, and she seeks a new mate to begin nesting again, thus being able to nest three times annually (a good survival strategy when food is plentiful). In addition to the natural climatic disasters, with which they have obviously coped during their evolution, the flightless cormorants are now also faced with the additional pressures of human disturbance, oil pollution, and the Japanese harvestors of sea cucumbers.

Steamer Ducks

Steamer ducks are large and very specialized diving ducks that use their feet and wings underwater, and also upend to feed in the shallows like the dabbling ducks. They are primitive birds, whose ancestry and relationship is still undetermined, although they are believed to be most closely related to the shelducks. They reminded sailors rounding Cape Horn bound for California of the side-wheeled Mississippi steamboats, which gave rise to their common name. Lowering their heads and necks until their backs are awash, they flap their wings and paddle furiously with their very large feet, thrashing and foaming the water and creating a wake as they move at speeds of up to fifteen knots. The locals call them loggerheads on account of their very large heads. In their large size, physical appearance, nesting habits, and marine habitat they resemble the northern eider ducks, but these similarities have resulted from convergence, not common ancestry. Steamer ducks are carnivorous and search the shoreline and kelp beds for molluscs and crustaceans, crushing the shells with their powerful bills. They are believed to mate for life and defend their territory viciously against intruders, being especially aggressive when raising their young; but their eggs and ducklings are still taken by skuas, gulls, and caracaras, and by foxes on the mainland. They have the most unusual calls, both sexes making a variety of grunts. There are two kinds of steamer ducks: the fliers and the flightless. The flying steamer duck is a bird of the coasts and islands of southern South America, from the Valdez Peninsula in Argentina to Concepcion in Chile, and also occurs on Tierra del Fuego and the Falkland Islands. It is the very obvious ancestor of the two flightless species, their differences being related solely to flight. It is a slightly smaller, less bulky bird with longer wings and is able to fly inland to freshwater lakes and rivers to nest, whereas the flightless species are naturally confined to the coastal waters and shorelines.

The flightless steamer ducks are restricted to segments of their flying ancestor's range. One occurs on the Falkland Islands, while the other coincides with the flying steamer duck on the South American mainland and its offshore islands. It seems a straightforward case of separate populations of flightless steamer ducks developing in those locations, to which their increasing flightlessness restricted them, while their full-winged ancestor continued to fly among them. However, there is now speculation that the flightless steamer ducks living on the Patagonian coast of Argentina may be the Falkland Island species, and not the Magellanic one, as always supposed. If this is so it raises the interesting question of why the same flightless duck should occur at two separate locations. The answer may lie in the fact that the Antarctic circumpolar current turns northward as it rounds Cape Horn and flows around the Falkland Islands and along the southeastern coast of South America. The implication is that the flightless steamer ducks first evolved in the

Falkland Islands, a typical environment for the development of flightlessness, except perhaps for the presence of the Falkland Island wolf.[1] They later reached the mainland on the current across 300 miles (482 km) of ocean, not an impossible task for a flightless marine bird, and then gave rise to the Magellanic flightless steamer duck on the Pacific Ocean coast.

The flightless steamer ducks actually have very powerful wings, but they are small and cannot be fully extended. Their bones and tendons are well developed and their pectoral muscles are at least as large as those of flying ducks of similar size. They are heavier than the flying steamer ducks, which has put more pressure on their wings when steaming, but they are also assisted by their very large feet, which measure 16 inches (40 cm) across when the webs are extended. One disadvantage of small wings is their reduced ability to brood their growing ducklings, which have to huddle together to keep warm. Hunting is apparently not a threat to the survival of either of the flightless steamer ducks, which are thick-skinned, tough, and not considered edible. The only major human predation occurs in the Beagle Channel, which separates the southern-most islands of Chile from Tierra del Fuego, where crab fishermen bait their traps with duck carcasses.

Magellanic Flightless Steamer Duck (*Tachyeres pteneres*)

This is the largest of the steamer ducks with a body length of 25 inches (64 cm), but although Captain Cook reported a specimen weighing 28 pounds (12.7 kg), the top recorded weights since then are only 13 pounds (5.9 kg) for a male and 8 pounds (3.6 kg) for a female. It is a very aggressive bird with a powerful head and heavy bill, and is basically pale gray, mottled with brown. Males have bright-yellow bills, whereas the females' bills are greenish-yellow, and they have browner heads than the drakes. This species is still common in coastal waters along rocky shorelines and in kelp beds on the tip of South America, ranging from Peninsula Valdez on the Patagonian coast at about 42°S latitude to slightly farther north on the Pacific Ocean coast. It also occurs on offshore islands including Tierra del Fuego, but avoids the eastern entrance to the Straits of Magellan, where there are strong tidal currents. The Magellanic flightless steamer duck nests near the shoreline, usually in grass but occasionally in woodland. Like the other flightless species, it lays up to eight eggs (but they differ in color, being ivory instead of buff) and has the same incubation period of thirty-five days. It was recently suggested that the northern population of this species, living along the shoreline of the province of Chubut, Argentina, should be considered a separate species—the white-headed steamer duck (*Tachyeres leucocephala*).

Falkland Flightless Steamer Duck (*Tachyeres brachypterus*)

This species is endemic to the Falkland Islands, an archipelago in the South Atlantic Ocean east of southern Argentina, comprising the main islands of East and

West Falkland and 200 smaller ones. They are all grassy islands with low shrubs, deeply indented coastlines, and many sandy beaches. The endemic flightless steamer duck, known locally as the logger duck, is a heavily built bird, although slightly smaller than the Magellanic species; its maximum recorded weight is 10 pounds (4.5 kg) for a male and 7 pounds (3.1 kg) for a female. It is also browner in overall coloration, because its pale feathers are bordered with maroon. Young birds and adult females have more brownish-red on the head and neck, while the mature males have a pale-gray head, which becomes white as they age. Both sexes have a white eye ring and orange feet. It is still a common bird, seen regularly in the harbor of the capital Port Stanley, and along the shore near the town. It feeds close to the shore among the rocks and in the kelp beds, and nests in the grass close to the shoreline, occasionally using abandoned penguin nests. The five to eight buff-colored eggs are laid between September and December, and are incubated solely by the female for approximately thirty-five days, although the male helps to raise the ducklings and protects them aggressively.

Brown Teal

Teal are small, surface feeding, dabbling ducks that are widely distributed throughout the world, especially in the Southern Hemisphere. The New Zealand brown teal, a brownish bird with green head and white eye-ring, is one of the rarest species. Once common, like so many of New Zealand's birds it was reduced to endangered status by a number of threats, including wetland drainage, shooting, alien predators, and the introduction of bird diseases. It has been the subject of an intensive breeding and reintroduction campaign by New Zealand's Department of Conservation, and has been returned to several parts of its former range. Brown teal are not enthusiastic fliers, and New Zealand's foremost ornithologist of the nineteenth century, magistrate Sir Walter Buller, wrote of the teals' reluctance to fly, but despite many years of isolation in the absence of predators, until recently at least, they never lost the ability to fly. However, those that were blown south toward the Antarctic long ago and settled on two sub-Antarctic island groups—the Campbells and the Aucklands—became flightless. Typical of remote, sea-girt islands, land predators were absent initially, there was ample food in the small freshwater pools and streams and along the shoreline, and the climate was cool and wet with no great extremes of temperature. There was simply no incentive to fly.

Auckland Island Flightless Teal (*Anas a. aucklandica*)

This small, brown flightless duck is restricted to the Auckland group of islands, which lie about 250 miles (400 km) south of New Zealand. Discovered in 1840 by naturalists aboard HMS *Erebus* and HMS *Terror*, commanded by Sir James Ross, it resembles the New Zealand brown teal, but lacks its white collar and the dark line on the back of the mantle. Its head and neck are dark grayish-brown, the nape is greenish and the mantle is dark brown vermiculated with black. Its back, rump,

Auckland Island Flightless Teal *This duck is descended from New Zealand brown teal, which were blown to the remote sub-Antarctic Auckland Islands long ago, and in the absence of predators lost their flight. It is mainly nocturnal and feeds in the kelp beds along the shoreline, moving inland to nest in vegetation adjoining streams and pools. Note its shortened wings with the primary feathers high on its sides.*
Photo: Courtesy Otorohanga Kiwi House

and tail are blackish-brown, the breast is dark chestnut, and the abdomen dark gray. The breeding male has a glossy green head and a narrow white collar. The Auckland Island flightless teal is still considered a race of its mainland ancestor, unlike the Campbell Island flightless teal.

It is mainly a bird of the shoreline where it feeds in the kelp beds, mostly after dark; and when it ventures inland it stays near water courses and ponds. It has very short wings and cannot fly, and when disturbed flaps across the water surface using its wings and feet for propulsion. It has never been seen to dive, and when threatened on land may hide in rabbit and petrel burrows. Vegetation near streams and ponds is the favored site for nesting and incubating their four tan-colored eggs. Skuas are thought to be the teals' only predators, as their remains have been found in skua middens. Although little is known of the teal's food preferences, it is believed to favor invertebrates, and the hardening of the gapes of its mouth is thought to be an adaptation for eating spiny isopods—small crustaceans resembling woodlice.

The Auckland Island flightless teal is already extinct on Auckland Island, and survives only on other islands in the group, namely, Adams, Disappointment, Rose, Ewing, Ocean, and Enderby. An attempt to relocate it early last century to Kapiti Island, northwest of Wellington, was unsuccessful. It is a rare bird, with a total population of between 1,000 and 1,500, but is now considered secure due to the remoteness and inaccessibility of its island home, and the fact that it has bred regularly at New Zealand's National Wildlife Center at Mt. Bruce, North Island. The London Zoo exhibited the first specimens seen in the western world at the end of the last century, but it has rarely been seen in collections outside New Zealand since then.

Campbell Island Flightless Teal (*Anas nesiotis*)

This flightless duck occurs only on the Campbell Islands, a group of small and steep tussock-covered islands lying between Antarctica and New Zealand, about 500 miles (804 km) from the latter and 250 miles (400 km) from the Auckland Islands. It was exterminated on Campbell Island, the main island in the group, by brown rats that came ashore from a visiting vessel in 1810. Although it was hoped that the duck may still survive on other islands in the group, the lack of sightings convinced biologists it was extinct, but in 1975 the New Zealand Wildlife Service discovered several specimens on Dent Island (a very steep islet of 56 acres (22 ha) just 1 mile (1.6 km) west of Campbell Island) where they had survived because of its freedom from introduced predators. Dent Island still contains the only known wild population of this race of the flightless brown teal, which currently numbers about seventy-five birds. Despite its small population it is thought to be relatively secure at present due to the island's remoteness and the difficulty of landing there. It lives in the tussock grassland from sea level up the steep slopes almost to the island's summit, preferring gullies with running water, boggy areas, and small pools. Its presence is often indicated by its habit of probing for food, its beak leaving distinctive holes in the peat moss. It is not known if it enters the sea, but like its relatives on the Auckland Islands it seeks the safety of petrel burrows when threatened. In addition to its small wild population, captive efforts to propagate this duck are now achieving success. Individuals were caught in the mid-1980s and again in 1990, and after years of effort at New Zealand's National Wildlife Center they finally began breeding in 1994. By 2000 the captive population had increased to sixty birds, and nine ducklings were raised in the latest breeding season on record—the summer of 2002/2003. Two groups of twelve birds each have been released on Whenua Hou (formerly Codfish) Island, with an 88 percent rate of establishment success achieved. A rat eradication program was conducted on Campbell Island in the winter of 2001, with poison baits spread by helicopter, and during a visit in 2003 the island was declared rat free. The decision was then made to reintroduce its native teal, the aim being to provide fifty captive-bred birds for release.

There has been disagreement over the classification of the Campbell Island flightless teal since specimens were first collected in 1886. It is a smaller, browner bird than the teal of the Auckland Islands, but is so similar that some taxonomists believed it may not even be a race or subspecies, but merely a fairly recent straggler from the Auckland Islands. Others thought that both insular flightless teal were races of the New Zealand brown teal. Modern science has finally superceded speculation, for recent blood analysis has proved that the Campbell Island flightless teal is indeed a distinct species, which obviously reached the islands long ago while still capable of flight.

Note

1. The Falkland Island wolf was discovered in 1690, and was still common when Charles Darwin visited the islands in 1833. It was killed by fur traders for its pelt and by sheep farmers to protect their flocks, and was extinct by 1876. It was possibly a feral domestic animal, like the dingo in Australia, taken to the islands by prehistoric people. It ate penguins and waterfowl, so this is another case of flightless birds developing in the presence of a predator, or at least holding their own, if it arrived after they had evolved.

7 Virtually Grounded

Birds with rudimentary wings like the kiwi and cassowary obviously cannot fly, whereas the albatross and condor, with their great wingspans, are clearly superb aerialists; but the case for flight or flightlessness is not always so obvious. Many birds are in an intermediate stage and fly rarely, and then only clumsily and weakly for short distances. Limited flying ability is less obvious in sedentary forest dwellers that do not require strong flight to move among the branches or even between neighboring trees, and in ground-dwelling birds that merely fly to the next patch of cover when they are flushed. However, the flying ability of several species is still uncertain, and some, such as the tapaculos, have been studied insufficiently to determine whether they have already lost the power of flight. Many other birds are reluctant fliers but can fly strongly when really pressed. Yet there is certainly a difference between reluctant flight and the inability to fly, including the early stages of loss of flight, which has resulted in semiflightlessness.

Reluctant fliers fly well enough when forced, but otherwise seldom bother, whereas semiflightless birds have an uncontrollable reason for flying poorly. This may be due to changes in their wing structure, their muscles and skeleton, or even in their physiology, which have occurred because of their unwillingness to fly. Some birds can only flutter short distances, often just a few yards, barely getting airborne for even such short flights. Should these species arbitrarily be considered flightless, since their limited wing-flapping ability hardly deserves to be called flight? On the other hand, if the ability to get airborne still exists, can birds really be considered flightless? The efforts of some species parallel our earliest attempts at powered flight, when even getting off the ground for 100 yards was considered flying; yet others can flap neither farther nor higher than a heavy domestic chicken. Others have never been seen to fly, but perhaps this is only because they have rarely been observed or studied anyway. It is difficult to know where to draw the line, and I have pondered long on this question, finally reaching the conclusion that the only certainty is total flightlessness. Birds that can still "hop-fly" like the early air-machines and those that flap heavily between neighboring trees have yet to totally lose the use of their wings for flapping flight and therefore cannot be considered flightless, but they are certainly semiflightless.

For several of these hop-fliers the reduction of their flying powers was certainly predictable, since their environment encouraged it, and it actually occurred in other related species. Yet the reverse is true of others that have almost lost the use of their wings: flightlessness seems improbable as they inhabit environments rich in predators. Birds in Madagascar, Australia, and Central America have reached their virtual flightless condition alongside carnivorous mammals, birds, and reptiles which were obviously a threat to them.

The rails, a family of shy, retiring birds, show the whole range of flying ability from strong intercontinental flight to complete flightlessness, with several intermediate forms whose flying capability is unclear. Resident sedentary species such as New Zealand's spotless crake and banded rail fly awkwardly and not very far; in fact the crake's longest recorded flight was just a few yards. But apart from the degree of awkwardness involved, this is typical rail behavior. Of greater interest is why they are not already totally flightless, evolving as they did on islands free of predatory land mammals. Some rails certainly seem more genetically inclined to losing their power of flight than others. For example, the weak-flying white-breasted or Cuvier's rail of Madagascar, an island rich in small predators, lives alongside the Madagascar rail, which still flies strongly. Rails that fly have one thing in common: they do so reluctantly, although this does not necessarily mean they are poor fliers. Also, reluctance has yet to compromise the flying ability of many resident tropical rails that do not migrate, and therefore have even less need to fly.

Other reluctant fliers in the bird kingdom include the kelp goose of Patagonia; the megapodes or mound-builders; and the hemipodes or button quail (not the Chinese painted quail that pet keepers now call buttonquail), all of which prefer to run when disturbed, and even when they fly as a last resort they do so poorly. The Asiatic pittas, and many of the 230 or so species of tropical American antbirds, are sedentary terrestrial birds whose rounded wings are more suited to short flights through the undergrowth than sustained flight in the open, but they are hardly semiflightless. However, not all reluctant fliers are skulking birds of the undergrowth. The roadrunner, for example, is adapted for running swiftly in the open on its long, powerful legs. It can certainly fly, although it seldom does. Large grassland birds like the bustards and guinea fowl also prefer to run as a first line of defense, but neither could be considered semiflightless.

Semiflightless birds have reached that stage in their evolution for many of the same reasons as the now totally flightless species, but have not regressed as far. They have little need to fly far or often, since they are mainly birds of the undergrowth that feed and nest on or near the ground. However, there are some surprising differences. Unlike the flightless birds (excluding the large ratites), many of the semiflightless species have evolved on continents where they were naturally exposed to numerous predators. Most live in tropical America, including the fifty species of tinamous and the fifty-six species of tapaculos. It comes as no surprise that the tinamous, ancestors of the ratites, are semiflightless. What is more surprising is that they did not give rise to other living flightless species in South America in addition to the rheas, and why they themselves have not totally lost their flight.

Like so many birds, semiflightless species were unable to cope with the unnatural environmental changes of the last few centuries, especially the loss of habitat and the arrival of introduced wild predators and feral, domesticated, farm, and pet animals. Several were exterminated and others are now quite rare. The Jamaican wood rail (*Amaurolimnas concolor*) preferred to run when threatened and flew only for short distances with very labored flight. It inhabited swamps, stream banks, and jungle undergrowth, but had no defense against introduced cats, rats, and finally the mongoose, and the last specimen was collected in 1881. The Chatham Island fernbird (*Bowdleria rufescens*), the most distinctive race of the fernbirds, with vivid chestnut back and spotted underparts, stood little chance against the burning of its habitat and the introduced cats and dogs, and was extinct by 1900. Four forms of the unique New Zealand wrens of the family *Acanthisittidae* were lost. All three races of the bushwren (*Xenicus longipes*) were exterminated in the twentieth century. The North Island bushwren (*Zenicus longipes stokesi*) was last seen in 1955, the Stewart Island bushwren (*Z. l. variabilis*) was extinct by 1965, and the South Island bushwren (*Zenicus l. longipes*) has not been seen since 1972. It is uncertain whether they were just poor fliers or were already flightless, but in the absence of proof of their flightlessness they are included here. The fourth wren lost was the Stephen Island wren (*Xenicus lyalli*), which is believed to have been flightless because it had a poorly developed keel, short rounded wings, and was never observed to fly. It was exterminated in 1894 by the lighthouse keeper's cat on its small island in the Cook Straits, and is included with the historically extinct species in Chapter 9. The huia (*Heteralocha acutirostris*), a most distinctive New Zealand bird and a poor flier, succumbed by 1908 to the demand for its silver-tipped black tail feathers, which were prized by the Maoris and then by commercial feather collectors and museums. The Seychelles warbler (*Acrocephalus sechellensis*), with its restricted insular range— just the Seychelle Islands—numbers about 2,000 individuals, after being brought back from the brink of exinction when its population was less than fifty birds. The survival of several tapaculos and tinamous is threatened by land clearance for agriculture; and like so many skulking and secretive birds of the undergrowth, their actual numbers and the survival strategies required to save them may not be determined until it is too late.

The situation of the semiflightless birds of the world raises some interesting questions. Will they, for instance, eventually lose their flight if allowed to continue their pre-twentieth-century existence? This seems unlikely, for the ability to continue their current, or at least former, lifestyle for any length of time appears rather doubtful for most species of wildlife. For birds that have obviously taken a very long time to reach their current poor level of flight, the passing of just another century, rather than a millennium, will only result in their demise or their complete control in small populations in protected areas, rather than changes in their flying ability. Then, if protected areas can remain inviolate long enough, will the reduced populations and the unavoidable inbreeding produce flightlessness as fast as it happened on island paradises with limited founder stock? Or will the increased use of their wings as a result of the threats to their survival result in improved flight? If the threats are that serious the survival of the species would be in question, not its improved ability to fly.

However, an idyllic life on an island free of predators does not necessarily result in loss of flight, and some insular birds are disinclined to fly, but this may not necessarily mean they cannot. Consider the case of the St. Helena plover or wirebird (*Charadrius sanctahelenae*). An obvious windblown descendant of the Kittlitz plover (*Charadrius pecuarius*) to its tiny mid-South Atlantic island from neighboring South Africa, although 1,000 miles (1,609 km) away, it has been on St. Helena for long enough to change sufficiently to be considered a distinct species. Yet despite being a ground-dweller, a ground-feeder, and a ground-nester, the St. Helena plover still flies strongly, although it is disinclined to do so unless really forced.

The Species

Tinamous

The tinamous are a very ancient group of birds, which are generally accepted to be the ancestors of the ratites, their relationship having been confirmed by DNA

Elegant Crested Tinamous *The tinamous are the ancestors of the flightless ratites. Although they are not themselves flightless, and still have a keeled sternum and flight muscles, their wings are short and rounded and they fly rarely and not very far. When startled they prefer to run on their strong legs, but they have a small heart and lungs in relation to their size and soon become exhausted.*
Photo: Courtesy Peter Nash

analysis, plus the similarities in the calcite nature of their eggshells and the structure of their paleognathous palates. The forty-seven known species are mostly rather dull birds with barred or mottled plumage, which resemble large quail and grouse, with a small head, a thin decurved bill, long neck, and heavy body; and the females are larger than the males.

Tinamous are widespread in the neotropics in a variety of habitat, including lowland and montane forest, scrub, grasslands, and semi-arid regions, but they are relatively little known due to their secretive habits. They are extremely shy birds, and the forest species are seldom seen and rarely photographed. They stay in the shelter of the dense undergrowth, and a quick glimpse of a small, chicken-like bird running across the path or skulking away into the bush is the usual view of a tinamou. Tinamous are very wary birds, wary of predators and recently of hunters, for they are a favorite game bird throughout their range. Those native to the more open grasslands and desert scrub of southeastern Brazil, and even farther south in the grasslands of Paraguay and Argentina, are more frequently seen. Attempts to introduce them to to several countries as a potential game bird failed with the exception of Easter Island, where the Chilean tinamou (*Nothoprocta perdicaria*) is now established.

The tinamous' wings are short and rounded, although they still have a carinate or keeled sternum and quite well-developed flight muscles, but they fly rarely and then only clumsily and with poor maneuverability. They also have thick, chicken-like legs, yet they cannot run far before becoming exhausted, and it is believed that this shortage of breath and poor flying ability may result from having a small heart and lungs in relation to their size. Yet tinamous have survived for millions of years, and it is surprising that they have not totally lost their flight. However, they do live in a rather hostile environment, with lots of potential predators in the form of birds, mammals, and reptiles; and with their poor flying ability other strategies were obviously necessary for survival. Their cryptic plumage and the ability to remain motionless are usually sufficient to avoid detection, but if that is insufficient they slink away into the undergrowth or even hide in a hole. Their next escape mechanism is to run, but they soon become exhausted; as a last resort they fly: low, clumsily and noisily, and not very far. With the exception of the members of the genus *Tinamus* the tinamous roost on the ground. They all forage on the ground for their food, and when adult are omnivorous, eating seeds, shoots, berries, and invertebrates such as insects, worms, and beetle larvae. Like the chicks of most omnivorous and seed-eating birds, young tinamous are intially quite insectivorous.

Outside the breeding season the forest species are mainly solitary, but the grassland tinamous may flock in large numbers—up to 100 elegant crested or martineta tinamous (*Endromia elegans*) have been seen together—but their breeding behavior is one of the most unusual of all birds, for they practice a form of polygamy known as polygyny. Male tinamous establish and defend their nesting territories, while the females band together in threes or fours and go looking for a suitable mate. After a brief display the male mates all the females, who each lay two or three eggs in his nest, and then go off to find another mate, leaving the male to incubate the eggs and raise the chicks on his own. Each female may lay eggs in several nests during the breeding season. The exception to this rule is the ornate tinamou (*Nothoprocta ornata*), which forms a pair bond for the duration of the nesting period and the parents share the incubation and chick-raising duties.

Tinamou eggs are the loveliest of all birds' eggs, being highly glossy in mono-chromatic colors such as purple, lavender, chocolate, and bluish-green. A clutch from several females may total twelve eggs, and their incubation period is 19 to 21 days. The chicks are precocial[1] and can run soon after hatching and fly when three weeks old—the only time tinamous really fly during their lifetime.

Little Tinamou (*Crypturellus soui*)

The male little tinamou is a dumpy brown bird with a dark-gray head, white chin, and dark-brown back and wings, while the female is similar but her underparts are rufus. It is one of the smallest species, with a wide range in both savannahs and humid forests from southern Mexico to southern Brazil, and also in Trinidad. It is rarely seen, but its low whistle is often heard. Its eggs are glossy lavender in color.

Tepui Tinamou (*Crypturellus ptaritepui*)

This is a very rare tinamou, restricted to the mountain forest on the tops and slopes of two tepuis or flat-topped mountains, Ptari-tepui and Sororopan-tepui, in interior Venezuela, at elevations from 4,265–5,900 feet (1,300–1,800 m). Its total range covers just 10 square miles (16 square km) and its survival is threatened by fire, which frequently occurs on the tepui's slopes.

Lesser Nothura (*Nothura minor*)

This species is also very rare; its numbers are believed to be no more than 10,000 and possibly as few as 2,500, fragmented over a large area of central and southeastern Brazil. It is one of the smallest tinamous, a plump, rufus-colored bird about 8 inches (20 cm) long, with back and wings blackish-brown barred with rufus, and pale-buff underparts that are spotted with dark brown on the breast and throat. Recent sightings have occurred mostly in protected areas where the habitat is undisturbed, but elsewhere it is at risk, as its natural habitat—the campo cerrado grasslands—is being rapidly converted to farmland. Even when hunted with dogs, it prefers to run to escape rather than fly.

Rails

Among the sedentary, nonmigratory species of rails there are many poor fliers. From the accounts of their behavior it appears that their flying ability stems as much as or more from inability than reluctance; and more detailed investigation of their lifestyle may indeed result in some being considered flightless. Several certainly show evidence of being emergent flightless birds, with soft flight feathers and shortened primaries, in addition to their reluctance to fly. Three of the numerous species that could be considered en route to flightlessness are described below.

Banded Rail (*Rallus phillipensis*)

Unusual for a rail, this species is a very attractive bird, clad in many shades of brown and with grayish underparts banded with black and white, a bright pale-orange patch on the breast, and brown above with white spots and it is about 12 inches (30 cm) long. Its head is dark rufus brown with a white stripe above the eye and a white chin. (See the color insert.) It is a bird of the marshes and the undergrowth along coastal shorelines, and is omnivorous, but prefers crustaceans and molluscs.

The banded rail is probably the greatest island colonizer of all birds, and is one of the most widespread of all the rails, ranging from India through Southeast Asia into Indonesia, Australia, and New Zealand. Over such a wide range, which includes many isolated islands, it is not surprising to find that it has many races or subspecies, most of which fly reluctantly and some that fly very poorly indeed, often just a short distance before dropping back into the undergrowth. The race that lives on New Caledonia and the Loyalty Islands, *R. p. swindellsi*, runs fast but seldom gets airborne. In New Zealand and the Auckland Islands, *R. p. assimilis* is a very reluctant flier, with long secondary wing feathers that almost match the length of its shortened primaries. The natural history and flying ability of several races, especially those confined to remote islands, are little known. This applies to the Cocos rail (*R. p. andrewsi*), of the eastern Indian Ocean's Cocos Keeling Islands (of which less than 1,000 survive after predation by feral cats and rats); the unnamed rails (*R. p. xerophilus*) of Goenong Api Island in the Banda Sea; and *R. p. christophori*, which occurs on San Cristobal Island in the Solomon Islands. Two races of the banded rail were exterminated as a result of settlement of their islands years ago. The flightless Dieffenbachs rail (*R. p. dieffenbachi*) of Chatham Island was extinct before the end of the nineteenth century due to predation by the settlers' cats and the burning of marshes to create farmland. The last MacQuarie Island rail (*R. p. macquariensis*) was seen in 1880, exterminated by cats and competition from the introduced weka, a larger and more aggressive rail.

Spotless Crake (*Porzana tabuensis*)

Like the banded rail this species is also a colonizer of remote islands, and has a wide range throughout Oceania and Australasia. It is believed to be the ancestor of the flightless rail of Henderson Island. It is a shy and elusive bird of the marshes and swamps, and in New Zealand prefers the dense beds of cattails (*Typha*) along the edge of both freshwater and brackish pools, rather than the flax-dominated wetlands. It is another rail that seldom takes wing, and the New Zealand race (*P. t. plumbea*), which is also the one that occurs in Australia and Tasmania, flies only a few yards before giving up and dropping back into the reeds. The spotless crake is one of the smaller rails, just 8 inches (20 cm) long and weighing little more than 2 ounces (57 g). It has a black bill and dark-brown head, back, and wings; bluish-gray underparts; and its undertail coverts are black barred with white. Its red eye-rings and reddish legs are its most colorful features. With its secretive habits and its active periods being mainly at dusk and dawn, it is more often heard than seen, and its compressed shape is perfect for weaving through the reed stems. It dives readily and is a good swimmer.

White-throated Rail (*Canirallus (Dryolimnas) cuvieri*)

A common bird in Madagascar, this rail is the ancestor of the flightless rails of nearby Aldabra Atoll, which still survives, and of Assumption Island, which is extinct. Although a ground-dweller, living on an island with several small carnivores, it can only flutter weakly for short distances. It is a medium-sized rail with a long red bill, dark-greenish upperparts, a rufus head and a startlingly white throat, and its undertail coverts are also white.

Mesites or Stilt Rails

The mesites are a family of little-known, thrush-sized birds endemic to Madagascar, of which three species are known. Two have been placed in the genus *Mesites* and one in *Monias*, and they are referred to mainly by these scientific names. They are a very ancient family of birds, whose relationship to other species is still uncertain, and they have been grouped at various times with the *Galliformes* (game birds) and with the *Gruiformes* (cranes and rails). They are believed to be remnants of a more widespread group of birds that survived only in Madagascar. The mesites are omnivorous and eat a wide range of invertebrates, plus seeds and berries. They are terrestrial, sedentary birds, with short but well-developed wings and long, full tails, but although their wings appear functional they are of little or no use for sustained flight because of the extensive degeneration of the breast muscles. Such a condition is unusual for small birds on an island with so many carnivorous animals, which include several species of mongoose and the larger cat-like fossa. None of these predators are large enough to have posed a threat to the now-extinct elephant birds.

Brown Mesite (*Mesitornis unicolor*)

This is a rare and secretive bird that is rarely seen and has never been observed flying. This may just reflect its reluctance to fly, or could be an indication that it is either in the early stages of losing its flight or is in fact already flightless. Its habitat is the lower mountain slopes of the humid evergreen forest of eastern Madagascar, which remains only as a very narrow strip running the length of the island from north to south. This bird has a very patchy distribution, and is considered vulnerable due to the decrease and fragmentation of its habitat. It occurs in several protected areas where it is secure, but elsewhere there are no controls on environmental degradation and its habitat is rapidly being destroyed. The brown mesite's survival depends entirely upon the preservation of the remaining undisturbed primary rain forest, but it is also threatened by dogs and rats near habitations.

The brown mesite is about 12 inches (30 cm) long, and has a small head and thick tail, with dark rufus-brown plumage on its back and wings, pale pinkish-brown underparts, and a variable white streak or spot behind the eye. The sexes are similar in adult plumage. They are omnivorous birds that search the leaf litter for invertebrates, turning the leaves over with their bills, and when infrequently seen they have been in pairs or small groups. Their nests are built just above the ground and only a single egg is laid, an unusual practice that will certainly not aid their rapid recovery if they can be adequately protected. Estimates of their numbers in recent years have varied from 2,500 to 10,000 birds, but all are agreed that the species is in serious decline.

White-breasted Mesite (*Mesitornis variegata*)

The white-breasted mesite is an inconspicuous, terrestrial rail-like bird, about 12 inches (30 cm) long, with a small head, plump body, and thick tail. Both sexes have similar rufus-brown plumage on their backs and wings, white breasts, and a pale-chestnut breast band. Their heads are striped with brown and chestnut. This mesite is a bird of the forests of northern and western Madagascar, where it has a restricted and localized distribution, the only major population being in the dry forest on the central western coast at Ankarafantsika. It is a purely terrestrial species, which searches for invertebrates and seeds in the leaf litter, and has never been observed in flight. The white-breasted mesite is declining in numbers and is classified as a vulnerable species due to its restricted range and the threats facing it, which include forest clearance for charcoal production and slash-and-burn agricultural practices, resulting in small and fragmented groups that are now quite isolated from each other. The latest estimates place its population at 8,000–10,000 birds.

Monias or Sub-desert Mesite (*Monias benschii*)

The monias is the most common mesite, an unusual, long-legged rail-like bird 13 inches (32 cm) long, with greenish-brown back and wings. The male has a white belly and a pale breast mottled with black, while the female has rufus-spotted underparts. Both have long, decurved red-and-black bills, long pinkish legs, and a full tail. The monias is restricted to the semidesert scrub and dry deciduous spiny forest in a small area of coastal southwestern Madagascar, between the Mangoky and Fiherenana rivers, where its survival is threatened by charcoal burners, hunters, and predation by dogs and introduced rats. It is a ground-dweller that prefers to run when alarmed, or may fly weakly for short distances or just to perch on a low branch.

The monias is the most social of the mesites, a gregarious species that is usually seen in flocks of up to ten birds. Like the other mesites it eats invertebrates and seeds, digging in the sand and overturning leaves to find its food. The most recent estimate of its population is 80,000 birds, and possibly as many as 100,000, making it by far the most plentiful of the mesites. Unfortunately, it is doubtful if any indigenous Madagascar bird can possibly be considered safe in the face of the increasing threat posed by deforestation and the encroachment of settlement, plus trapping and predation by dogs and rats.

Hoatzin

The hoatzin (*Opisthocomus hoazin*) is a bronzy-olive bird with chestnut under-parts and a pale breast. It is the size of a small chicken, and in some respects resembles one, with a spiky red crest, bright red eyes surrounded by bare blue skin, a long neck, chicken-like body, and a long black tail. It is 25 inches (63 cm) in length and weighs almost 2 pounds (900 g); it has such an unpleasant, musky odor that it is not hunted for food, but its eggs are eaten by humans and monkeys. Its home is the tropical northern half of South America, in a large area of lowland rain forest in the drainage basins of the Amazon and Orinoco rivers, always close to water. It prefers the still or slow-moving waters of oxbow lakes, lagoons, and swamps.

Hoatzin   *The hoatzin is a bird of the swamps and riverbanks of the Orinoco and Amazon river basins. It has large wings but is quite sedentary and flies only short distances between trees; this lack of activity has resulted in weak breast muscles. In addition, it has a very large crop, in which it bacterially ferments its diet of swamp plant leaves; this has further compromised the space needed for flight musculature.*
Photo: Photos.com

The hoatzin is a very primitive bird of uncertain ancestry, its most unusual feature being the claws on the chick's wings, which help them to clamber about in the branches around the nest within a couple of days of hatching, and to climb out of the water. These claws drop off as the chicks mature, but they still climb very awkwardly and damage their growing wing feathers. When alarmed, they drop into the water and swim beneath the surface using their feet and wings. They have anisodactyl feet, in which the first digit (the hallux) points backward and the other three toes point forward, which is the standard for birds.[2]

The claws, reminiscent of the first feathered reptile *Archaeopteryx*, have puzzled scientists since the hoatzin's discovery, and have caused indecision about its ancestry. It has been linked with a number of birds including the pheasants, the cuckoos, and the touracos, but is currently considered most closely related to the game birds, *Galliformes*, where it occupies a suborder of its own. Unlike the other game birds, however, whose chicks leave the nest soon after hatching and which have wing feathers when they hatch or within a few days, hoatzin chicks are naked at first but soon grow down feathers, and stay in or near the nest for several weeks as the growth of their wing feathers is delayed.

Hoatzins are sedentary birds that rarely move far from the colony where they hatched. They are most active at dusk and dawn, and hide in the undergrowth or the dense foliage of trees bordering rivers and swamps during daylight. Their food consists mainly of the leaves of marsh plants, especially arum, but occasionally they also eat fish and crabs, and are the only known birds to ferment food in their crops. Like the ruminant mammals and other leaf- and grass-eating birds, they cannot digest their food without assistance, which is provided by symbiotic bacteria which ferment and break down the indigestible plant fiber into digestible sugars. Unlike other birds—such as the grazing geese and the heather-browsing grouse—which break down vegetation in the caecum, the hoatzin ferments food in its large sectional crop, which as in all birds is an extension of the esophagous, normally used for storing food. The mash is then regurgitated for the chicks, or swallowed, and enzymes in the stomach then neutralize the volatile organic plant compounds such as alkaloids and terpenes, which plants produce to protect their leaves. With a full crop the hoatzin is top-heavy and when perching must lean on its breastbone to prevent overbalancing. The hoatzin is not a good flier. It has large, broad wings but they are weak because of their rarely used flight muscles, whose size has also been further compromised by the space needed for the large crop. Although it can glide between trees it flies very clumsily and only for short distances, perhaps just across the lagoon to feed, and then crash-lands into the foliage.

Hoatzins are social birds that live in family groups or small colonies of up to fifty birds, made obvious by the noisy grunting and wheezing that are their only vocalizations. They may be polygamous, and the members of the family group assist with the incubation and chick raising. They breed all year, although mostly in July and August just when the rains begin, and nest close to each other in bushes and low trees over water, building a rough platform of twigs on which they lay two or three eggs, which are incubated for twenty-eight days.

Henderson Island Fruit Dove

The Henderson Island fruit dove (*Ptilinopus insularis*) is restricted to Henderson Island, a World Heritage Site in the southeastern Pacific Ocean, lying 107 miles (172 km) northeast of the more famous Pitcairn Island, where the mutineers from HMS *Bounty* settled. It is a small, raised coral atoll edged with steep cliffs of bare limestone, and is mostly densely vegetated with scrub and low trees. It is almost surrounded by a fringing reef which has two narrow access channels for boats. Its ecosytems are still practically undisturbed and it is considered the best remaining example in the world of a raised coral ecosystem. Its isolation has encouraged the evolution of a unique fauna and flora, with ten endemic species of plants and four birds, one being a totally flightless rail, another a fruit dove that has almost lost its flight.

Doves are just small pigeons, and the island's endemic fruit dove, a plump little bird about 9 inches (23 cm) long, has pretty gray and yellowish-green plumage with a dark-red crown, but unlike all other living wild pigeons is semiflightless. It is still arboreal, however, climbing in the branches in search of fruit and berries, and flapping heavily from tree to tree, its short wings making longer, sustained flight impossible. It certainly cannot migrate between islands like many of the Oceanic fruit pigeons. Its population has been estimated at about 3,000 birds.

Henderson Island was originally settled by the itinerant Polynesians, but has been uninhabited since they left at the end of the fifteenth century, and is now a port of call for Pitcairners and yachtsmen who collect wood, fruit, and coconuts. Its fruit pigeon was fortunate, however, as it was obviously considered too small for the pot; the island's other pigeons, the larger *Ducula pacifica* and *Ducula galeata*, were both exterminated by the Polynesians, and the bones of a large, flightless pigeon have also been found in kitchen middens.

These plump, colorful doves prove that even such fast-flying treetop-dwelling species, which in the South Pacific often fly long distances between islets in search of fruiting trees, can still lose their flight when the environment does not require them to fly. This should hardly be a surprise however, since the most famous extinct bird of all—the dodo—and its relatives, the solitaires, were descendants of pigeons.

Tapaculos

The tapaculos are a relatively little-known group of birds, of about fifty-six species, that live in South and Central America. Most, if not all, are in various stages of losing their flight, and some may already be flightless. Their wings are weak, short, and rounded and the keel is virtually absent, so even those that still fly do so weakly. This is an unusual condition for birds living on a continent that has so many snakes, lizards, hawks, and small mammals able to prey upon them.

Tapaculos are related to the ant thrushes and are mostly small, dull birds ranging in size from a wren to a thrush. They are all secretive, fast-running ground-dwellers with long legs and sturdy feet, and they hold their tails cocked-up. They live in a variety of habitat, from the high-altitude scrub forest to the dense undergrowth of the lowland rain forest, and also in the dry pampas where they hide among bushes and in tall grass. Most species are ground-nesters, and some use burrows, while a few nest close to the ground in trees. As a result of their skulking behavior and dense habitat they are seldom seen and little known, and five new species have been described in the past decade. They are often called feathered mice as they scurry for cover when disturbed rather than fly.

Grey Gallito (*Rhinocrypta lanceolata*)

The grey gallito lives in the grassy plains of the Argentinian pampas, where it nests in thickets. It is one of the larger species, about 8 inches (20 cm) in length and has grayish-olive upperparts, grayish-white belly, and a crested rust-colored head.

Bahia Tapaculo (*Scytalopus psychopompus*)

This is almost certainly the rarest species of tapaculo, known from only three museum specimens collected in small areas of fragmented coastal forest in Bahia State in southern Brazil, and with an estimated wild population of about fifty birds, although it has not been seen recently. A slaty-gray bird with white underparts and with rufus on its flanks, it is one of the smallest species, barely 5 inches (13 cm) long. Destruction of the Atlantic coastal forest has led to its perilous position.

Brasilia Tapaculo (*Scytalopus novacapitalis*)

Another tiny species, not quite 5 inches (13 cm) long, this tapaculo is a white-and-gray bird with a dark plum-colored head and rufus rump and vent. It lives in swampy riverbank forest and dense undergrowth adjoining streams in central Brazil, and is now a very rare bird, its habitat having suffered considerable damage from the burning of neighboring grasslands for farming.

Lyrebirds

The lyrebirds resemble large pheasants, with small heads, long necks, large legs and toes, and with a length of 36 inches (91 cm), of which the tail takes up more than half. They are one of the largest members of the *Passeriformes*—the perching birds. They are shy and retiring birds, and with their small, rounded wings and degenerate pectoral muscles, their flight is restricted to short and rather clumsy efforts between close trees, and rarely more than a few yards, but they also make long downhill glides. To gain height for roosting they hop from branch to branch, and then glide back to the ground in the morning.

Lyrebirds are fast runners, and prefer to run when threatened; they also use their powerful feet to scratch the soil away in their search for invertebrates. They are restricted to Australia and are distantly related to the scrubbirds. They are so ancient they may be a relict species of Gondwanaland origin like the emu, cassowary, and possibly the scrubbirds, unlike Australia's other bird fauna, which came later from the north. If so, they have been isolated on the continent for millions of years. Lyrebirds are great mimics, and most of their calls are borrowed from other sources—other birds, dogs barking, and even vehicle engines—coupled with their own whistles and shrieks. They are mainly insectivorous, and scratch in the leaf mould like the pheasants for invertebrates, frogs, and small lizards.

Lyrebirds were killed for their plumes in the nineteenth century, but have been strictly protected for many years. However, their low numbers early in the last century alarmed biologists and twenty-two birds were introduced into Tasmania in the 1930s and 1940s, as insurance against their possible extinction on the Australian mainland. Superb lyrebirds caught in Victoria were released in Mt. Field National Park from where they have since spread to occupy a large area of the southwestern forests. The birds released originally have multiplied to a current population of over 8,000.

Superb Lyrebird (*Menura novae-hollandiae*)

This species lives in the wet eucalyptus and beech forest east of the Great Dividing Range from southern Victoria to southeastern Queensland. It is still a common bird although it is now threatened by forest clearance in some areas. It is not a colorful bird, having dark grayish-brown upperparts and paler underparts, and its only bright feathers are the reddish ones on its throat, but the male's ornate tail is spectacular and more than compensates for the dull plumage. It contains sixteen feathers, the outermost ones resembling an ancient Greek lyre when spread in display. The lyre-shaped feathers are black above and pale gray or white below; they are broad and curled and enclose a dozen filamentous central feathers and two long, tapered and curled plumes. The female lyrebirds are smaller and have similar dull plumage but lack the lyre-shaped display feathers.

The male lyrebird builds several display mounds, sometimes as many as a dozen, by scratching earth into a low pile, and uses each one in turn when he displays to attract females. Standing in the center, he spreads his tail up and backward over his head, singing clearly at the same time. The females attracted by this performance are mated, usually several by each male, and go off to lay their eggs and raise the chicks, unaided by the male. Their nest is a woven, dome-shaped pile of roots and sticks, usually on the ground, with an inner chamber lined with moss and feathers. A single egg is laid and incubated for forty-one days and the chick stays in the nest for six weeks. It matures slowly and males do not attain their full tail plumage until they are at least three years old.

Albert's Lyrebird (*Menura alberti*)

Albert's lyrebird has a much smaller range, just the subtropical rain forest of northeastern New South Wales and southeastern Queensland. Forest clearance is also the main threat to its survival, although it is still relatively common. This species is smaller and darker than the superb lyrebird, its plumage a rich chestnut, with paler underparts. Males lack the lyre-shaped feathers, and instead have filamentous central tail feathers and fan-shaped outer ones that are black above and gray below. This species also makes a display "mound," but of trampled vegetation, and displays in a similar fashion to the superb lyrebird.

Scrubbirds

The scrubbirds are a very old Australian bird family with just two species, the noisy scrubbird and the rufus scrubbird, small birds resembling long-tailed wrens. Their closest relatives are the lyrebirds, and like them they are possibly of direct Gondwanaland origin. Scrubbirds are sedentary and very secretive ground-dwellers, occurring only in isolated populations in dense undergrowth, where there is a deep layer of leaf litter that contains invertebrates such as beetle larvae, worms, and small snails, which form their diet. Both are cryptically colored birds, and when alarmed they either hide in the dense undergrowth or run very fast on their short, sturdy legs. They have rudimentary clavicles, a reduced sternum, and short rounded wings, and their "flight" is consequently limited to fluttering just above ground for a few yards. They are rarely seen and the only way to estimate their numbers has been to count the singing males, which have a loud, clear voice.

Although their territories may cover 25 acres (10 ha), most of an individual scrubbird's activities are usually concentrated in a 5-acre (2 ha) area, and they nest in the same place each year. Both species are now rare, due to loss of habitat to logging, agriculture, drought, and bush fires. They are also very vulnerable to predators, especially feral cats, as they nest close to the ground. The scrubbird's display is similar to the lyrebird's, with fanned tail, dropped wings, and quivering body, coupled with singing and mimicry. For such small birds they have a very powerful voice and their loud calls, which are reminiscent of a whipcrack, can be heard from some distance through the forest.

Rufus Scrubbird (*Atrichornis rufescens*)

This species lives in the thick undergrowth of isolated montane rain forest in northern New South Wales and southern Queensland, where there are believed to be about 2,500 breeding pairs, surviving mostly in protected areas such as Mount Lamington National Park. The rufus scrubbird is about 8 inches (20 cm) long, and males are rufus brown above, barred with black, and have a white throat, black breast, and rufus abdomen. The female is similar but lacks the black breast.

Noisy Scrubbird (*Atrichornis clamosus*)

This species is similar but slightly larger than the rufus scrubbird, the males being brown above, barred with black, and having white, black, and buff underparts.

The females lack the black upper breast. The noisy scrubbird lives in the thick undergrowth of the heath and scrub bush of the extreme southwestern corner of Australia, and was thought extinct for several decades, a victim of the bush fires that destroyed its habitat. A few birds were rediscovered in 1961 in a 20-mile (32 km) stretch of dense scrubland on the south coast of Western Australia. Within a few years this population, based on the number of singing males, was about 100. Insect eaters, the noisy scrubbirds had not suffered directly from the bush fires, but the fires had destroyed the deep leaf litter in which they foraged for invertebrates, and it has since been determined that it takes almost ten years to regenerate the depth of leaf litter needed to support this invertebrate life. The measures taken to protect this small population included fire control programs and the translocation of birds. The objectives of the recovery plan established by the government of Western Australia were to achieve and maintain a population of more than 300 singing males and their cohorts in the Albany Management Zone, and to establish new populations in a zone west of Albany. This program was a great success, and the noisy scrubbird now has a total population in excess of 1,500 birds.

Wattlebirds

The wattlebirds are another ancient group of birds whose origins and relationship to other species are still uncertain, although they may be descended from the Australian bell-magpies and apostle birds. The common ancestor of these birds is likely to have been isolated on New Zealand when Gondwanaland broke up about 100 million years ago.

Wattlebirds have short, weak wings and very labored flight over short distances, but they mostly hop around on the forest floor searching for invertebrates, or climb trees to launch themselves into a glide. They nest low to the ground and have therefore been very vulnerable to introduced predators, especially cats and stoats. There were three species of wattlebirds until almost a century ago. One of them, the crow-sized huia (*Heterolocha acutirostris*), had a restricted range in the southern mountain forests of North Island, and was already scarce when the first European settlers arrived, as the Maoris prized its tail feathers. With the rapid loss of habitat and overhunting by museum collectors it was extinct by 1908. The other species—the saddlebacks and the wattled crows—barely survive, due only to the continuing efforts of the Department of Conservation.

North Island Saddleback (*Philesturnus carunculatus rufusater*)

These saddlebacks are dark greenish-black birds with a conspicuous chestnut "saddle" on their backs (actually their wing coverts and mantle) and a chestnut rump. There is a small wattle on each side of the throat, which varies in color from yellow to reddish-orange, and males have larger wattles than the females. They are ground-dwellers, which search the forest floor for invertebrates and fallen fruit. It is a slightly smaller bird than the race from South Island and has a narrow, yellowish-buff line along the front edge of its saddle, which is missing in the other race. Young birds resemble their parents in coloration.

They were common when European colonization began, but suffered greatly from forest clearance and the loss of their habitat, plus predation by rats, cats, stoats, and ferrets. At the beginning of the last century this race was restricted to Hen Island in Hauraki Gulf, but birds have been transferred from there to a number of other predator-free islands including others in the nearby Chicken group, and to Motukawanui Island in the Bay of Islands. The North Island saddleback now numbers about 2,000 birds.

South Island Saddleback (*Philesturnus c. carunculatus*)

This is a truly unique bird, in which the young have cocoa-brown plumage up to the age of fifteen months, totally different from their parents, a condition which is quite unusual among birds. At the beginning of the last century this race survived only on three islands—Big South Cape, Pukeweka, and Soloman—just off the larger Stewart Island. Rats from a shipwreck colonized these islands in the early 1960s and exterminated two other endemic birds. To save the saddlebacks the Wildlife Department (now the Department of Conservation) translocated thirty-six birds in 1964 to other predator-free islands and other translocations were made as their populations grew. These included twenty-five birds transferred in 1984 to Motuara Island in Queen Charlotte Sound, and sixty from Big South Cape Island to Breaksea Island in Fiordland in 1992. The South Island saddleback is now established on ten predator-free islands and although still a very rare bird, with a population of about 700, it has responded to the conservation efforts and its survival appears assured. Its official recovery plan calls for 4,000 birds to be eventually established on South Island's rat-free offshore islands.

North Island Kokako or Wattled Crow (*Callaeus cinerea wilsoni*)

Kokakos are very attractive bluish-gray birds about 15 inches (38 cm) long, of which there are, or were until recently, two races that differ only in the color of their wattles. The North Island race has blue wattles at the base of its bill, and the South Island race has orange wattles. Kokakos are forest birds, which were common when the first European settlers arrived in New Zealand at the end of the eighteenth century, but were lost as their forest was destroyed and alien predators such as black rats, brush-tailed possums, stoats, and feral cats began their systematic destruction of the native birds. Like the other wattlebirds, they have short and rounded wings, and their flight is feeble and limited to short distances, so they depend more on their sturdy legs to escape potential threats. In keeping with their mainly terrestrial habits they nest on the ground or close to it, so their eggs, chicks, and incubating females were vulnerable to predation. The outcome was a pre-dominantly male population and an even greater reduction in breeding success.

The North Island kokako is the only race to survive on the New Zealand mainland. It is a bird of mature hardwood podocarp forests, and has been the subject of considerable preservation efforts by the Department of Conservation, whose recovery plan ambitiously calls for its reestablishment in as much of its former habitat as possible. It still survives in remnants of native forest in North Island, especially in Te Urewera National Park and in several locations in the Bay of Plenty. In 1984 the Forest Service bought the 1,000-acre (400 ha) Aislabies Forest near Rotorua, which is an important habitat for the bird. Kokakos are now estab-lished on Little Barrier Island from translocations made in the 1980s, and from there some were moved to Kapiti Island and then to Tiritiri Matangi Island in Hauraki Gulf in 1998. The current population is believed to be about 2,000 birds, but if it is still skewed in favor of males, as suspected, then there could be only 400 breeding pairs.

The South Island kokako (*Callaeus c. cinerea*) was a common but localized bird in South Island at the end of the nineteenth century, but declined rapidly and was last seen in 1967. It is hoped that some may still survive in remote regions of Fiordland and Stewart Island.

Fernbird

The fernbird (*Bowdleria punctata*) is endemic to New Zealand and several of its neighboring islands, where it inhabits scrubland, heath, gorse, and ferns in open country, and the dense vegetation of freshwater and salt marshes, swamps, and low-lying river banks. On Snares Islands it lives in the tussock grass, the only similar habitat available. It is a small bird, about 7 inches (18 cm) long, with rich brown upper plumage and white underparts that are streaked and spotted with dark brown. Its head is chestnut-brown with a white eye stripe. The wings and tail are also brown, and its tail always appears frayed.

It is a secretive bird that is rarely seen, but gives away its presence with its squeaky calls, especially in spring when it is more territorial. The fernbird is a feeble flier for two reasons. Its wings are poorly developed and it has very loose plumage due to the lack of barbs to hold the feathers together. It flies only when hard-pressed, although very low and noisily, and seldom more than 20 yards (18 m). In the open, it is easily run down and caught, but in the reed beds it threads its way through the stems like a rail and quickly vanishes. It is insectivorous, and eats a wide range of invertebrates, including spiders, flies, moths, caterpillars, beetles, and their larvae.

It was a common species throughout the islands until the middle of the nine-teenth century, when the loss of habitat and introduced predators reduced its numbers. It is now localized due to its specific habitat preferences and the loss of such land to drainage and agriculture, and is therefore vulnerable to disasters. The fires in 1984 and 1989, which destroyed over half the vegetation in the Whanga-marino Wetland, a Ramsar site of 20,000 acres (8,000 ha) in the Waikato Basin of North Island, destroyed one-third of its 9,000 fernbirds. The species is therefore included in long-term survival plans, and thirteen fernbirds were transferred in May 2002 to the predator-free island of Tiritiri Matangi in the Hauraki Gulf. They have already bred and by the spring of 2004 the population there numbered about thirty birds.

There are five surviving races or subspecies of the fernbird. Those on North, South, Stewart, and Whenua Hou Islands have only slight plumage differences, whereas the race on Snares Island is less distinctly streaked and has broader tail feathers. A sixth subspecies, endemic to Pitt and Mangere Islands in the Chatham group, was extinct by 1900 due to habitat loss, overgrazing by introduced goats and rabbits, and predation by introduced cats.

Seychelles Warbler

The Seychelles warbler (*Acrocephalus seychellensis*) is one of the world's rarest birds, restricted to the Seychelle Islands in the Indian Ocean north of Madagascar. Just over 5 inches (12 cm) long, it has dull olive-brown upper plumage and pale buffy-yellow underparts, with a long bill and blue-gray legs. Pairs bond for the whole year, and usually lay only one egg, but they make up for this by breeding twice and even three times annually when food is plentiful. The Seychelles warbler still flies, but very poorly. It was originally confined to Cousin Island, a granitic outcrop in the western Indian Ocean, with a land surface of 70 acres (28 ha), and by 1965 its numbers had dropped to just fifty birds, due mainly to the clearance of the island's scrub vegetation and the dominant lettuce tree (*Pisonia grandis*) for growing coconuts, plus predation by the introduced barn owl. In 1968, when the island was purchased by the International Council for Bird Preservation (now BirdLife International), there were only thirty warblers left, but the conservation measures taken resulted in the most spectacular bird preservation success.

The island was granted Nature Reserve status by the Seychelles government, the barn owls were removed, and the natural vegetation was encouraged to re-generate. The warblers are monitored and research into their genetic variation and parentage is currently carried out. Aride Island is also now managed as a nature reserve. On Cousin Island the warbler's numbers had increased to 323 in 1997 which appears to be about the island's holding capacity. Twenty-nine birds were transferred to neighboring Aride (168 acres/68 ha) in 1988 where they had in-creased to 1,600 by 1997. A second group of twenty-nine birds from Cousin was released on Cousin Island (75 acres/30 ha) in 1990. The population of the three islands in 2004 was about 2,000 warblers and fifty-eight were transferred from Cousin to Denis Island to start another breeding group.

Central American Wren Thrush

The Central American wren thrush (*Zeledonia coronata*) is a small bird that is so unique it occupies a family of its own within the very large group of American wood warblers. It lives only in the wet mountain forests of Costa Rica and Panama, especially in bamboo thickets and along stream banks. A plump, short-tailed, long-legged bird, barely 5 inches (13 cm) long, the wren thrush is mainly dark olive and slate-gray but has a bright cinnamon-orange crown. It is quite terrestrial and spends its time on the ground, hopping about in the undergrowth, flicking its wings as it searches for insects, and it also nests on the ground. The wren thrush has short wings, reduced flight muscles, and its keel is virtually absent; it seems unlikely that it can fly with such modifications. In fact, it has never been seen to flutter more than a yard or two at a time, and may well deserve to be considered flightless. The wren thrush runs fast on its long legs when threatened, which has obviously provided some protection in an environment rich in terrestrial snakes and small mammalian predators.

New Zealand Wrens

New Zealand's wrens are very ancient birds, whose closest relatives are some primitive South American passerines and Australia's lyrebirds and scrubbirds, all of

Rock Wren   *A tiny, almost tail-less bird, the semiflightless rock wren lives at high elevations in New Zealand's Southern Alps. In winter it stays in the alpine regions, surviving in rock cavities beneath the snow where it searches for invertebrates. It can only flutter short distances on its degenerated wings.*
Photo: Courtesy Department of Conservation, New Zealand. Crown Copyright.
Photographer: Rod Morris, 1975

which have a distinct tendancy toward flightlessness. Restricted to New Zealand, these tiny birds are believed to be descended from one of its earliest bird colonizations, perhaps in the Tertiary Period about 30 million years ago. After so long in the world's major flightless bird paradise it is surprising that they are not already flightless. There are only two living species, the rifleman and the rock wren, and both flutter weakly for short distances. Four other forms, actually all races of the bush wren, are also now extinct. One of these—the Stephen Island bush wren—was believed to be flightless and is therefore included in Chapter 9, but it is unknown whether the others were just poor fliers or had already also lost their flight. These small birds are insectivorous, and are sedentary ground-dwellers or tree-trunk creepers that nest in holes in tree trunks or in rock crevices.

Rock Wren (*Zenicus gilviventris*)

A unique New Zealand species, believed to have been on the islands since the evolution of the moas many millions of years ago, the rock wren is a tiny bird, barely 4 inches (10 cm) long, and virtually tail-less. It has dull green plumage above and brown below, with a creamy-white stripe over the eye. Unlike the four extinct races of bush wrens, it is not a bird of the bush, but one adapted for life at higher elevations where it was safer. A truly alpine and subalpine species, it is restricted to the mountains of South Island, where it hops and flutters in search of insects among the rocks and scrub at elevations between 3,000 feet (915 m) and 8,000 feet (2,450 m). It does not migrate down the slopes in winter, but stays on the heights where it survives beneath the snow in cavities between the boulders and in rock crevices. It can only flutter a few yards on its weak wings, and quickly hides among the rocks when threatened. The rock wren occurs on high mountain terrain from northwest Nelson down through the Southern Alps into Fiordland. It is generally considered to be monotypic, and a subspecies described from Fiordland, *Zenicus gilviventris rineyi*, is of doubtful status. Its population is very fragmented and localized, however, and it is now quite rare everywhere except in Fiordland, where it continues to decline due to the depredations of the stoats, which are initially attracted above the tree level by the mice living there. Even if the adult birds can avoid the stoats, their eggs and chicks in nests among the tussock grass or in a rock crevice are very vulnerable.

Rifleman (*Acanthisitta chloris*)

The rifleman is New Zealand's smallest endemic bird, being only 3 inches (8 cm) long. The male is bright greenish-yellow above with a white eye-stripe and white underparts. The hen is more brownish above and is slightly larger. It is more plentiful then the rock wren, and still occurs on all three major islands (where it is common in the higher altitude beech forests), except for the northern parts of North Island, and rarely moves beyond its small territory. The rifleman is now also established in man-made habitats, including managed pine forests. It also thrives on Great Barrier Island and Little Barrier Island in Hauraki Gulf, and has been translocated to Tiritiri Matangi Island. Some taxonomists have accorded subspecific status to the populations of North Island, South Island, and the Southern Alps, but this is not generally accepted. Like the rock wren, the rifleman's powers of flight are very feeble and it seldom flutters far, but it is still quite arboreal and "runs" up the trunk like a tree-creeper, searching for insects on leaves and in the bark. After reaching a height of about 30 feet (9 m) it "floats" down again on open wings. This species probably deserves to be considered flightless, in which case it would be the world's smallest flightless bird. (See the color insert.)

Notes

1. Precocial chicks are down-covered, their eyes are open, and they leave the nest soon after hatching. In contrast, altricial chicks are naked and helpless when they hatch, and remain in the nest until fledged.

2. Other birds—including the parrots, cuckoos, and woodpeckers—have zygodactylous feet, with two toes facing the front and two backward.

8 Briefly Grounded

Feathers have two main purposes: protection and flight. Body feathers protect and insulate the bird and assist in maintaining its temperature, whereas the larger wing and tail feathers are concerned with flight. But feathers wear out and are normally moulted annually and replaced with new ones. The number of feathers shed at one time varies, but most birds have a typical or sequential moult, where feathers are lost a few at a time so that their flying ability, insulation, and body temperature are not compromised. However, some have evolved the unusual behavior of moulting all their flight feathers at once and becoming temporarily and seasonally flightless until the new ones grow. It is not a common form of moulting strategy, as the risk of predation is obviously increased at such times. This temporary flightlessness occurs in several families of mostly aquatic species such as the waterfowl, grebes, loons, coots, and auks, which are flightless for three to five weeks, depending on the species. Temporary loss of flight is also an aspect of the unusual nesting behavior of the hornbills, and even shorter-term flightlessness, just for a few hours or perhaps a whole day, also afflicts several birds, not through the loss of feathers but because they have overindulged and cannot get airborne.

There is something very tempting about a bird that cannot fly. Even the appearance of flightlessness is used to great effect by many birds, including some that have been flightless for many millennia, such as the rhea and ostrich, which still practice the broken-wing display to entice potential threats away from their nest or chicks. Wild predators, domestic cats, and children are fascinated by a bird that appears to have difficulty flying and therefore appears easy prey. Consequently, periods of flightlessness, however short, are times of extreme danger for birds that normally fly. After all, if there was little risk at such a time they would probably have lost their flight permanently. The only birds that are relatively safe while flightless are the hornbills, protected in their tree-nest cavity from attack by snakes and monkeys, but the others must seek a safe haven for moulting, and must avoid drawing attention to themselves. This is why it is mostly aquatic birds that are temporarily flightless, for they have access to the relative safety of large bodies of open water during this very vulnerable period of their life cycle.

Waterfowl

With just a few exceptions waterfowl moult their flight feathers simultaneously and undergo a completely flightless period of up to four weeks while awaiting the growth of their replacement. Until recently it was thought that the six species of South American sheldgeese, such as the ashy-headed goose (*Chloephaga polioce-phala*) and ruddy-headed goose (*Chloephaga rubidiceps*) did not moult all their primaries at once, but observations in the wild have cast doubt on this, and they may also be flightless during the moult. Consequently, it is likely that there is only one member of the order *Anseriformes*—the ducks, geese, and swans—that does not moult all its wing feathers at once, and that is the magpie goose (*Anseranas semi-palmatus*) of Australasia. This atypical behavior, however, is understandable for a bird that begins its moult at the end of the breeding season when the seasonal swamps where it nested are drying out. Losing its power of flight at such a time would prevent it from migrating to open water and expose it to greatly increased predation from dingos and large monitor lizards.

Spring and fall in the Northern Hemisphere, when large numbers of waterfowl band together to migrate, is not the only time these birds congregate in large numbers. The biennial migration of birds northward to nest and then the reverse southward for the winter is a well-known phenomenon, but a lesser-known aspect of bird movement, especially of the seaducks, is the long migratory flight that many make for their post-breeding season moult, when they shed all their wing feathers. These flights are usually to a specific traditional site during the summer months in the Northern Hemisphere, with the birds generally flying north for this purpose, usually in very large flocks.

These moulting sites are important areas for the survival of many aquatic birds, as important as their summer breeding grounds, their wintering places, and the staging points between the two; they are therefore key areas for conservation that are vulnerable to predation, disturbance, and oil spills. Safety and food availability are the most important considerations for these seasonally flightless birds. The inability to fly in an environment containing predators naturally makes a bird more vulnerable than normal, and the waterfowl that moult all their flight feathers at once are no exception. As the colorful breeding plumage of some male ducks would make them more noticeable and susceptible to predation during their flightless period, many reduce their visibility with a post-nuptial moult (toward the end of the breeding season) prior to moulting their flight feathers. Drakes replace their colorful breeding plumage with an "eclipse plumage," at which time they resemble the duller females. For most of their flightless period they are therefore cryptically plumaged, although they may begin the body moult reversing this before they are fully flighted again.

Even though their dull plumage may not attract attention, they must find a safe haven for when they are relatively helpless, where predators and people cannot take advantage of their situation. They generally seek the safety of the open sea, large lakes, or offshore islands where terrestrial predators cannot reach them, and this chosen site is often used regularly by a population. However, birds anxious to moult their wing feathers do not just settle on the first large body of water they see, but follow definite migratory patterns to a traditional moulting site, which may be a long way from their nesting places.

Waterfowl migrate from the Canadian Maritimes to Labrador and southwest Greenland to moult, and geese that nest in the Arctic may fly to even higher latitudes for safety. From Arctic Canada king eider ducks (*Somateria spectabilis*) fly to western Greenland to moult in remote fiords and are flightless for at least three weeks. Denmark's National Environmental Research Institute has monitored these moult-ing congregations for several years and estimates that several thousand birds assemble at the moulting site. However, some of these birds, especially the true seaducks, such as the king eider duck, harlequin duck (*Histrionicus histrionicus*), and red-breasted merganser (*Mergus serrator*), leave the females to incubate the eggs while they take off to the moulting site. But the females have evolved a strategy of their own, and in Iceland many female king eiders abandon their nests when their eggs hatch and leave their ducklings under the care of other females who gather them into groups and act as chaperones, shep-herding them to the water to feed. European shelducks (*Tadorna ta-dorna*), which fly from western Eur-ope and the British Isles to traditional moulting sites in the North Sea, es-pecially Germany's Helgoland Bight, act in a similar manner, leaving their ducklings in the care of nonbreeding young adult ducks that have retained their wing feathers, when they fly off to moult.

It is not only the seaducks or coastal waterfowl that congregate in large numbers to moult. The inland freshwater ducks have similar behav-ior but generally do not fly such long distances, although these birds rarely moult in their nesting areas. Lidcliff

European Shelduck   *Leaving their ducklings in the care of nonbreeding females, European shelducks fly from their breeding grounds along the coast of western Europe to the North Sea where they are safe from predation while they are flightless. Soon after arrival they shed all their primaries and are grounded for several weeks until they regrow.*
Photo: Ernesto Lopez Albert, Shutterstock.com

Marshes, an 1,800-acre complex of wetlands in Manitoba, is a major moulting area for redhead ducks (*Aythya americana*) and canvasback ducks (*Aythya valisneria*). On Rutland Water in Leicestershire, the moulting flocks of tufted ducks (*Aythya fuligula*) have numbered 5,000 birds. While safety is a prime concern for flightless waterfowl, the choice of moulting site is also influenced by the availabilty of sufficient food for the whole flightless period, especially if they cannot swim to other bodies of water. Researchers on Lake Myvatn in northern Iceland monitored the correlation between the moulting diving ducks there and the lake's supply of aquatic insect life. They discovered that the number of flightless Barrow's goldeneye (*Bucelapha islandica*) on the lake was dependent upon the insects available, and that lowered levels of food when the birds arrived resulted in them going out to sea to moult.

Waterfowl that are terrestrial feeders and graze on grasses, such as the grey lag goose (*Anser anser*), need more than just a large expanse of water as they must come ashore to feed, so the availablity of grass at or near the water's edge obviously influences their choice of moulting site. But they rarely venture inland, preferring to graze on the grasses of the salt marshes and sand dunes not far from the water's edge, for a quick escape if threatened by predators. Unless an environmental disaster changes the nature of the site, moulting waterfowl return to the same place each year, and this regularity of use has allowed biologists to track and monitor their populations and movements. Moulting waterfowl are also easier to catch, and biologists take advantage of this for officially sanctioned banding purposes. Unfortunately, however, biologists with capture permits are not the only ones to take advantage of the ease with which flightless waterfowl can be caught, and humans have always been a very serious predator of these birds.

New Zealand's ducks were an important source of food for the Maori when the European settlers arrived, and Sir Walter Buller, in his *History of the Birds of New Zealand*, published in 1888, relates how they captured moulting grey duck (*Anas superciliosa*), scaup (*Aythya novaeseelandiae*), and shoveler (*Anas rhynchotis*) in large numbers in the Bay of Plenty district. Women and children in canoes drove the flightless ducks to the shoreline where they attempted to hide in the scrub and sedges but were caught by the hunters' dogs. Buller says 7,000 were taken in three days on one lake alone. Even now the opportunity to catch birds that cannot fly is seldom missed around the world. Moulting ducks and geese are "herded" into corrals in large numbers for food, a method now copied by bird banders. Eskimos in Siberia still catch thousands of moulting king eider ducks in this way each year. It is not only the large flocks that are vulnerable; the solitary species are by no means safe, even in regions such as Oceania, which is very poor in waterfowl. The only ducks I saw during several weeks collecting birds on remote Niuafo'ou in the Friendly Islands were the pair of Pacific black ducks, which my Tongan companions relentlessly chased across the island's central caldera lake. Although dabbling ducks, they did not flap across the surface but dove repeatedly, using both their almost featherless wings and feet to swim underwater, but were soon exhausted and caught by hand. On the Venezuelan llanos—the central seasonally flooded grasslands—social waterfowl such as whistling or tree ducks of the genus *Dendrocygna* are caught by the peasants in large numbers when they are seasonally flightless, and solitary South American species like the wild Muscovy duck are just as vulnerable. On Guyana's Upper Kamarang River I accompanied Akawaio Indians who hunted temporarily flightless wild Muscovy ducks, which stood little chance against their barbed arrows.

Other Waterbirds

In addition to the waterfowl several other groups of aquatic birds also shed all their flight feathers at once and are briefly flightless. They include the grebes, razorbills, guillemots, divers or loons, shearwaters, diving petrels, and at least one of the jacanas. The jacanas are totally freshwater birds, and the grebes and loons are mostly freshwater species, but the others are pelagic birds that live on the open seas and come ashore only to nest.

The Species

Grebes

Grebes are almost totally aquatic duck-like birds with long thin necks, pointed bills, and rudimentary tails. Their oil glands are well developed and they have very dense, waterproof plumage with up to 20,000 feathers each. Most of their food is caught underwater, yet they seldom remain submerged for more than a minute. Diving from the surface or sinking with barely a ripple, they swim powerfully with their feet, which are situated at the rear of the body, an adaptation for diving and underwater swimming. They also have stiff flaps or lobes on their toes which increase the surface area and aid swimming, and flexible tarsometatarsal joints between their legs and toes, which increases maneuverability, both in the water and in flight. However, they are awkward on land and seldom leave the water.

Grebes have small wings and are weak fliers, but like the rails, several species make long migratory flights twice annually. They are mostly freshwater birds, although in temperate zones they winter along the coasts. They have also shown that under the right conditions they can lose their powers of flight permanently; two of the twenty-one species are flightless (see Chapter 6) and two others became extinct recently (see Chapter 9). But even the flying grebes become completely flightless for up to four weeks when the nesting season has ended, when they gather in large groups in inland waters or close to the seashore and moult all their primary feathers at once. In British Columbia's Boundary Bay, 3,000 western grebes (*Aechmophorus occidentalis*) gather annually for their summer moult.

Alcids

Several members of the family *Alcidae*, which includes such diving seabirds as the auks, auklets, guillemots or murres, murrelets, razorbill, and the puffins, and which are known collectively as alcids, undergo a complete wing-feather moult and are flightless for several weeks. They moult at sea, which is where they spend their whole lives other than the short period on land when nesting.

Puffins (*Fratercula arctica* and *Fratercula corniculata*)

The puffins, of which there are several species, including the Atlantic puffin (see the color insert) and the horned puffin, are birds of the northern oceans. They are unmistakable birds with their large colorful bills, stout bodies, short wings, and red or orange webbed feet. They nest underground in burrows, about 3 feet (92 cm) deep, which they scratch out themselves with their feet on grassy cliff-tops high above the sea, but where soil is scarce they lay their single egg on a cliff ledge. Puffins can catch a number of fish without dropping those already in their beak, lining them up cross-wise before heading back to the colony. The puffins have short, narrow and pointed wings which they beat fast, almost in a whirr. These are perfect for both long distance flight and for "flight" underwater fast enough to chase and catch fish; their small size reduces drag but powers the bird like miniature feathered paddles. However, small wings cannot generate lift at low speeds, so a rapid vertical take-off from the water is impossible and puffins "skitter" across the surface to become airborne, often flying straight through the wave tops before they lift off. Getting aloft is easier from their cliff-top nesting sites as they launch themselves straight out into space. They can splash down on water easily enough but they stall as they slow down to land at their cliff-top nest sites and hit the ground with a thump.

Recent research into the activities of seabirds in the North Sea, off the northern coast of the British Isles, prompted by concerns over the extraction of oil and gas and the risk of oil spills, discovered that adult Atlantic puffins were flightless in the pre-breeding season of March and April, due to the loss of all their flight feathers. Although this loss obviously curtails their flying, it does not seriously impede their diving and swimming ability, as they compensate for the reduction in wing surface by more frequent flapping of their featherless wings, and move just as quickly underwater.

Razorbills (*Alca torda*)

Razorbills are large black-and-white alcids with a white line from the eye to the base of their stout black-and-white bill, which resembles an old-fashioned cut-throat razor. They are at home in the cold waters of the North Atlantic and southern Arctic Oceans, where they spend almost their whole lives, only coming ashore to breed. They are also colonial birds that nest in large colonies, but unlike the burrowing puffin, they lay their single egg directly onto the bare rock of a high cliff ledge. When the razorbill chicks are about three weeks old and still flightless, they drop down to the sea from their high nest ledges, and during the next two months they reach adult size and grow their feathers while at sea. Their parents accompany

Brunnich's Guillemots *These alcids, also known as thick-billed murres, nest high on cliff ledges. When the chicks are three weeks old and still lacking their flight feathers they free-fall down to the sea, where they are gathered together and escorted out to sea by the male birds. They moult their primaries then, and are flightless for almost four weeks, by which time the youngsters can also fly.*
Photo: Tony Hathaway, Dreamstime.com

them and while chaperoning them undergo a complete wing moult and are flightless for about four weeks.

Guillemots

The guillemots have similar unusual chick-fledging and adult-moulting behavior. At the age of three weeks their chicks are fully feathered except for their flight feathers, which have yet to develop. At dusk they free-fall from the cliff-tops 100 feet or more into the sea, where they are gathered together and escorted away from the shore by the male birds of the colony. These chaperones also undergo a complete moult of their primary and secondary feathers and, like the chicks, are unable to fly for about four weeks. In the species known as Brunnich's guillemot or the thick-billed murre (*Uria lomvia*) of the cold northern seas, the colony's chick-fledging and dropping into the sea is synchronized, with the earlier-hatched young awaiting the later hatchlings before all leap together off the ledges and try to maintain their balance with their stubby wings as they drop to the water. From the colonies on Baffin Island these flightless young and adult males then all migrate by swimming to their wintering grounds off the coast of Newfoundland and western

Greenland, a distance of about 600 miles (965 km). No other bird migrates as far at such an early age. This once plentiful bird is now seriously declining throughout its range. A nesting colony in north Greenland, which contained 500,000 birds in the mid-twentieth century, is now down to about 10,000. The guillemots also occasionally suffer from a weakness of the feather vanes, resulting in their rapid wear and affecting the birds' ability to fly until after their next moult.

Loons

The loons or divers are specialized aquatic fish-eating birds with heavy, sleek bodies, short tails, thick necks, and sharp, pointed bills. They are clumsy on land as their legs are set far back on their bodies, and like the puffins they take to the air with some difficulty, but are strong fliers once airborne. There are only five species of loons, all of which occur only in the Northern Hemisphere, their distribution just reaching the tropics. Their eerie, wailing calls are a familiar sound in the northlands, and they migrate south from northern lakes in the fall to coastal harbors and bays in more southerly latitudes that remain free of ice. Loons propel themselves with their feet, and eat mainly fish, caught after an underwater chase. They are excellent divers and like the penguins have good vision; their eyes can focus both underwater and in the air. Loons moult in late summer and are flightless for at least four weeks until they regrow their flight feathers.

Tubenoses

The birds known collectively as tubenoses are members of the order *Procellariiformes*, which have an extension of their nostrils in the form of an open-ended tube on either side of the upper bill. This excretes excess salt, preventing a buildup in the bird's body and allowing it to survive without ever drinking fresh water. With their connection to well-developed olfactory lobes, these birds are thought to have a good sense of smell, which helps them find food on the open ocean and locate their nest burrows in the colony. Two groups of tubenoses, the shearwaters and the diving petrels, experience the annual loss of all their flight feathers at the same time.

Shearwaters

The shearwaters are a group of seventeen species of small to medium-sized narrow-winged seabirds, which occur in all the world's oceans, and nest in huge colonies from sea level to high on mountainsides. They are nocturnal and return to their nest sites under cover of darkness. Most shearwaters are sociable birds, which gather at sea after the breeding season in large flocks or "rafts" of several million individuals (in the case of the common and wide-ranging sooty shearwater), during which time they moult their primary feathers and are flightless for several weeks.

Shearwaters feed on the surface or catch fish after an underwater pursuit following a dive from the surface or the air. Vast colonies of these nocturnal seabirds nest on the coasts and offshore islands of southwestern Australia, Tasmania, and New Zealand. Their gregarious nesting habits have not escaped attention, and they have been collected from their burrows in large numbers since the days of the visiting sealers and shipwrecked or marooned sailors.

The shearwaters arrive back at their colonies in September and October, and find their mate and their burrow. They pair for life, which may be fifteen years, and nest in the same burrow each year. On arrival they clean out their burrows, and the female lays a single egg. The eggs hatch in January and the chicks grow rapidly and are soon double the size of their parents. In early April the parents abandon their fat chicks, which are called squabs, and go to sea to moult. The squabs at that time are covered with down but do not have their wing feathers, and during the next three to four weeks they live off their fat stores and lose weight rapidly, while at the same time are surprisingly able to grow their feathers, which are made of pure protein. They then take off for the long migration north to Alaska where they spend the northern summer.

Many of the plump squabs, actually several hundred thousand of them, never make it to the sea, as they are still collected each year for human consumption. The chicks of two species of shearwaters, which are called muttonbirds, have been harvested for food and income since the first settlers arrived in Australia and New Zealand, and even before then they saved shipwrecked mariners from starvation and were a major source of food for the aborigines and Maoris. Early in the last century they were canned and marketed as "squab in aspic," and they can still be bought in food stores in Australia and New Zealand. The sooty shearwater (*Puffinus griseus*) is the muttonbird of New Zealand, with 500,000 squabs taken annually from their burrows on the islands off Stewart Island, where their traditional harvesting is the exclusive right of the Rakiura Maori. The muttonbird of Australia is the short-tailed shearwater (*Puffinus tenuirostris*), which lives in huge colonies. There are over 20 million mutton birds in almost 300 colonies around the coast of southeastern Australia and the offshore islands, especially the islands of the Furneaux Group, between northeast Tasmania and Victoria. The largest colony containing 3 million birds is on Babel Island off the east coast of Flinders Island. The short-tailed shearwater migrates almost 10,000 miles (16,000 km) twice annually, north up the western Pacific Ocean to the Arctic for the northern summer, and then back home down the center of the Pacific Ocean to breed during the Australian summer; a journey that takes about six weeks each way.

Diving Petrels

The diving petrels are small and stocky, short-billed and short-winged seabirds, which pursue fish underwater like the auks. Although all the species are black above and white below, and therefore resemble the auks of the Northern Hemisphere, they are confined to the southern oceans, and spend most of the year at sea. These birds have a similar role in the Southern Hemisphere as the auks—the puffins and razorbills—north of the equator. They make shallow dives for crustaceans and fish, rarely more than 2 feet (60 cm) deep, unlike their relatives the storm petrels, which take their food off the sea's surface. Like the shearwaters they are nocturnal and only come ashore at night to nest and feed their young. They are seldom airborne for long, usually flying short distances over the surface and through wave crests, before diving at full flight beneath the surface, using their wings to propel them underwater. When the breeding season is over, they moult all their flight feathers simultaneously when far out at sea. This flightless period lasts about four weeks, but they are still able to swim well enough underwater to catch their food.

Jacanas

The jacanas or lilytrotters have very long toes and claws, which spread their weight and enable them to walk on waterlily leaves. They are sedentary birds with slow, labored flight over short distances. The African jacana (*Actophilornis africana*) is a social bird with polyandrous breeding habits, in which the females lay several clutches of eggs each season after being mated by different males. Each male then incubates a clutch of eggs, and raises the chicks himself, carrying them under his wings for their first two weeks to keep them dry. After the breeding season the African jacana moults all its flight feathers at once and is flightless for several weeks; but this condition has not been observed in the other species.

Hornbills

Female hornbills undergo their annual moult and lose all their flight feathers while nesting, which may seem a rather risky practice, but the reverse is true as their nesting behavior generally ensures their safety. The hens of all species, except the ground hornbills, are sealed into their nest cavity with a mixture of mud, feces, and regurgitated fruit, which dries like cement, and keeps them secure from marauding snakes and arboreal mammals such as monkeys, civets, and genets.

The hornbills are an ancient family of birds that evolved millions of years ago in Europe, but are now restricted to Africa and Asia, extending west to the Philippines and New Guinea. They are dependent upon trees, and live in a range of treed habitat from wooded savannah and scrubbush bordering riverine forests in southern and eastern Africa to the mature tropical forests of West Africa, India, and Borneo. The rain forest species (which form 75 percent of the 54 living hornbills) are almost totally frugivorous,[1] but add frogs and lizards to their chicks' diet. The forest hornbills are very important to the ecology of the rain forest, as they act as dispersers of seeds, and some trees rely almost exclusively on them for their seed dispersal. The African arid-land species are smaller birds that are mainly insectivorous.

The hornbills are mostly large noisy birds, with the cavity-nesting species ranging in size from the great Indian hornbill (*Buceros bicornis*) and the helmeted hornbill (*Buceros vigil*), which reach 50 inches (1.27 m) in length, to the red-billed hornbill (*Tockus erythrorhynchus*) of the African wooded grasslands, which is just 16 inches (40 cm) long. They all have large bills, topped with a casque in most species, which is made of keratin and is not as heavy as it appears. The casque is hollow in all except the helmeted hornbill, whose solid casque is called hornbill ivory, and has been carved into intricate miniature scenes; but few pieces of this art exist.

Hornbills have traditionally been used for ceremonial purposes by the native peoples in many parts of their range. Borneo's Iban hunt the large rhinoceros hornbill (*Buceros rhinoceros*) for its black-and-white tail feathers, and in the Indian state of Arunachal Pradesh the Nishi tribal group have long used the bill and casque of the great Indian hornbill in their ceremonial headgear. As laws now prevent the killing of these birds, the Nishis have accepted fiberglass beaks so that their old customs are not lost.

Hornbills are unique in several ways. They are the only birds in which the first two vertebrae (the axis and the atlas) are fused, and in which the kidney has two lobes, all other birds having three lobes. It is in their nesting behavior, however, that the most obvious of their unique characteristics is apparent, for the female is imprisoned in the nest cavity for several months. As the hornbill beak is not adapted for drilling, even into soft wood, they cannot excavate their own nest holes, but depend upon woodpeckers and the natural wear and tear of age, storms, and lightning to produce tree cavities suitable for their nests. The cavity must provide sufficient space for the incubating hen and eventually her chicks as well, but size is not the only consideration; an entrance hole through which they can just squeeze is a very important factor, for if it is too large it will be impossible to seal.

Mate selection in birds is a very serious business, but is especially so in the case of the hornbills, for the male must convince his prospective partner that he is conscientious and quite able to support her and the chicks while they are sealed in the cavity. In all species that have been studied, the male is the one who finds a suitable nest hole and then invites the female to inspect it. The courtship display follows; he brings her fruit and in her presence continually checks the cavity to show his dedication. If she accepts his advances, mating then occurs and the female enters the cavity. With a mixture of clay, rotten wood, feces, and regurgitated fruit provided by the male, she then seals the entrance hole with his help from the outside, except for a narrow slit, just wide enough to poke food through. It takes up to a week to seal her in, and the mixture dries very hard and provides adequate protection from predators. The imprisoned female then lays, incubates, and raises the chicks.

For the duration of the incubation period, which varies from twenty-three to forty-five days depending upon the species, plus all or most of the fledging period (varying from fifty to ninety days), the hen hornbill is fed by the male through the slit. (See the color insert.) In the larger species she is therefore confined for at least four months. About two weeks after being sealed in, the female moults all her tail and wing feathers, and although safe from natural predators, bird collectors who felled the trees to reach a nest once posed the greatest risk to the confined birds. For a long time it was believed that the inability of the male to feed his family due to sickness or death also posed a great risk to nesting hornbills. This was despite the report of John Whitehead (who collected birds in Borneo in the 1880s and was responsible for the first prolonged survey of the fauna of Mt. Kinabalu) that he saw five bushy-crested hornbills (*Anorrhinus galeritus*), both males and females, feeding an imprisoned hen and her chick. Recent observations have now proved that in at least one-third of all hornbills, and possibly more, the male has assistance in feeding the incarcerated female and her chicks. His helpers are the grown young from the previous year and possibly other unpaired adult males. The female hornbill usually breaks out a week or two before her chicks, who seal the entrance up again. By then she has regrown her flight feathers and helps the male with the feeding.

Gluttons

The subject of temporary flightlessness in adult birds would not be complete without mention of the very temporary grounding which results not from the loss of the flight feathers, but simply as a result of overeating. This condition is called hyperphagia—just as it is in humans—for which the standard definition is

Lappet-faced Vultures *Vultures have evolved to gorge when the opportunity arises (perhaps quite infrequently), and they become temporarily grounded after a very large meal. The African species like the lappet-faced vultures feeding on zebra carcass (above) are then vulnerable to large predators such as lions and hyenas.*
Photo: Photos.com

"the excessive ingestion of food beyond what is needed to meet an animal's basic energy requirements," in short, gluttony. In some birds this results in short-term grounding.

Many birds overeat at certain times. Even hummingbirds do, and they may almost double their weight during a period of intense feeding, but in their case it does not compromise their flying ability. Just the reverse is true, in fact, for the purpose is to store the energy needed for their long migratory flights, like the ruby-throated hummingbird's nonstop crossing of the Gulf of Mexico. Other birds eat in excess simply because food is suddenly available after perhaps several days of enforced abstinence, which is the normal state of affairs for scavengers such as the vultures and condors, who rarely find food on a daily basis. When it is available they gorge and may eat several pounds of carrion at one sitting, sometimes increasing their body weight so much they cannot get airborne. Andean condors (*Vultur gryphus*) have been run down and caught when absolutely gorged with food, especially when feeding on the sea lion carcasses on the beach, as they do in Peru where the Andean heights are very close to the coast. Condors have also been captured in the Andes by placing a dead llama within a walled compound, from which they could not run to get airborne, and they certainly could not manage a vertical take-off. In Africa lions and hyenas have seized slow-moving vultures that flapped about trying to get airborne after gorging on a kill. Caracaras, long-legged New World falcons that rely mostly on scavenging for their food, are attracted to seabird nesting colonies where there is always plenty of carrion in the form of discarded eggs and dead chicks. Forster's caracara (*Phalcoboenas australis*), of the Falkland Islands and islands off the coast of Tierra del Fuego, have been observed so gorged at penguin colonies that they were unable to fly.

As the extent of knowledge of birds and their behavior increases it is likely that other examples of temporary flightlessness will become apparent. In birds such as the rails, most of which seldom fly even though they can, temporary flightlessness may also occur more frequently than appreciated. For example, a slate-breasted rail collected in New Guinea's Central Highlands, at first believed to be a new flightless species, was simply unable to fly because it had moulted all its secondary and primary feathers together; but whether this was just an individual circumstance or a typical and regular occurrence in the species as a whole is unknown.

Note

1. Except for some members of the genus *Tockus* such as the African crowned hornbill (*T. alboterminatus*) and the African pied hornbill (*T. fasciatus*).

9 An Ill Wind

During the millions of years that passed as birds were slowly evolving from the dinosaurs to the now familiar creatures of our gardens, woods, lakes, and seashores, many became extinct as they were superseded by others whose adaptations better equipped them to compete and survive. Many of these ancient birds were flightless, but rock impressions or fossilized bones are all that remain of their passing; so their shape and color, or even the length of their tails, is open to speculation. These long-extinct birds are considered prehistoric; the World Conservation Union's Species Survival Commission considers the year 1600 to be the borderline for historic extinctions. There is a good reason for this, since with very few exceptions, such as the moas that are depicted in Maori rock drawings, there is no record of the actual appearance of birds that died out before then, and modern illustrations of those species are merely artists' impressions based on bones or fossils.

Flightless birds have disappeared at a much faster rate in the last four centuries (the Historic Period) than during the evolution of birds, but in most cases complete museum specimens, eyewitness accounts, and illustrations are available, and their form, color, and feather structure are documented. At least fifty forms of flightless birds are known to have become extinct since 1600, including all the members of several families—New Zealand's moas, Madagascar's elephant birds, and the dodo and solitaire of the Mascarene Islands. It is these historically extinct birds that are the subject of this chapter, but even the current list is obviously incomplete, for the appearance of some recently lost species was never recorded and it is certainly possible that the remains of others have yet to be discovered. The partially fossilized bones of an extinct flightless ibis were found in a lava cave on Maui in 1974, and it is thought that the species may have survived until early in the nineteenth century. A large flightless waterhen is also believed to have lived in the Virgin Islands and Puerto Rico until the end of that century. Some sparse accounts of extinct flightless birds have been impossible to verify, and this includes three ratites—Levaillant's ostrich of North Africa, the dwarf rhea of southern South America, and the Chatham Islands kiwi—which are therefore not considered valid species. Then there is the "Lisianski Island rail," said by Russian sailors in 1828 to be the same as the rail of neighboring Laysan Island, but it was neither collected nor described scientifically and disappeared even before the Laysan rail.

Exploration, settlement, and the development of agriculture and the industrialized society have been the major threats to birds, and the flightless species suffered more than those that could fly. Surprisingly, in most species this was not primarily due to their flightlessness, but because they lived on islands or once-isolated lakes that were more vulnerable to change than continental land masses and confined the birds to their fate. Many island birds capable of flight were also lost.

Most forms of plant and insect life, and all the land birds, arrived on remote islands on the wind, but the same winds that were the making of the islands biologically were eventually responsible for their downfall. The very winds that blew the ancestors of the present-day flightless birds to their oceanic paradises helped mankind cross the great oceans, which until then had been a barrier to all terrestrial mammals. Homo sapiens was the first land mammal other than bats to reach remote islands, and the bird life suffered accordingly. With the exception of the Syrian ostrich (*Struthio camelus syriacus*), and the Atitlan grebe (*Podilymnus gigas*) and Colombian grebe (*Podiceps andinus*) on their remote mountain lakes, all the flightless birds that were exterminated historically lived on islands.

The great increase in bird extinctions began with the advent of world exploration and settlement, beginning almost 2,000 years ago with the colonization of islands in the central and south Pacific Ocean by the seafaring Polynesians, who exterminated many insular birds. The carbon-dating of bones from the Hawaiian Islands has shown that two-thirds of the native birds disappeared between the settlement of the islands by the Polynesians and the arrival of European settlers early in the nineteenth century. These extinctions included seven species of flightless rails and several flightless geese and ibis. New Zealand was populated by many species of giant moas, which were ratites and therefore related to the present-day cursorial kiwis, and in the largest species reached a height of 10 feet (3 m) and weighed about 450 pounds (204 kg). The belief that the Polynesians exterminated all the moas and several other flightless birds, including a giant goose, is supported by the fact that piles of bones of many species of moas and thirteen other extinct birds have been found at the campsites and middens of the early Maoris. The Easter Island rail, the eastern-most flightless bird of the South Pacific islands, plus all the other endemic species on the island, were exterminated by the Polynesian settlers, who eventually brought about their own demise through overpopulation and deforestation. The date of the rail's demise is unknown but is likely to have been prior to the Historic Period.

Many years later the Polynesians were followed across the Pacific by the navigators, Magellan, Drake, Bougainville, and Cook; and then by sealers, whalers, and settlers, which resulted in the second great wave of bird extinctions. Island fauna suffered severely from the seventeenth century onward as a result of exploration and colonization. Fortunately, descriptions, illustrations, and whole specimens exist to show what most of those birds looked like. Although the birds of the Pacific Ocean islands suffered most during this period, islands in the middle of both the Atlantic and Indian Oceans also lost several endemic flightless birds. In northern seas the great auk (*Pinguius impennis*) and the spectacled cormorant (*Phalacrocorax perpsicillatus*), both island-dwellers, were slaughtered to extinction. Practically all the losses of flightless birds occurred on islands simply because that was where most of them had evolved.

Island birds suffered from a variety of threats, and a combination of assaults was generally involved in their extinction. The Polynesians cleared the forest for agriculture; took pigs, dogs, and chickens with them to many islands, and Polynesian rats (*Rattus exulans*) came ashore from their canoes. The ships of the navigators, sealers, whalers, and convicts also carried rats and mice to distant shores, and islands were deliberately stocked with goats, sheep, and rabbits as future sources of food. The European settlers who followed them were mostly farmers, and their domesticated farm and pet animals accompanied them. Then they introduced predators such as stoats, weasels, and hedgehogs, and allowed their cats and dogs to go wild and live off the land. These animals and the all-consuming omnivorous domestic pig, which became feral on several islands, posed very serious threats to the native birds, and the results of their depredations were compounded by the loss of habitat as land was cleared for settlement and farming. Having survived the long sea journey, successfully colonized their new home, and changed their ways to suit its conditions, island birds could not cope with such sudden and drastic changes, which in comparison to their evolution occurred almost overnight.

However, despite all these new and combined threats to their continued existence, it was two small rodents that had the most serious effect on insular birds. On many oceanic islands, birds suffered more from the predations of the black or tree rat (*Rattus rattus*) and the brown or Norway rat (*Rattus norvegicus*) than from any other introduced pest. These highly carnivorous and very aggressive rodents were responsible for the extinction of many birds and have been implicated in the loss of others. The black rat has been more destructive generally to bird life on islands because of its tree-climbing ability, but the brown rat has taken a heavier toll of ground-nesting birds, both flightless and flying. In addition, the Polynesian rat is now known to be far more destructive to bird life than was previously thought. On Lord Howe Island, 500 miles (805 km) off the coast of New South Wales, the endemic fantail, flycatcher, blackbird, and white eye, all flying birds, were extinct by 1930 as a result of the arrival of black rats via a shipwreck in 1918. Surprisingly, the flightless Lord Howe Island rail (*Gallirallus sylvestris*) survived, although barely.

Attempts to eradicate rats from islands have been very expensive and generally unsuccessful, but the recent clearance of rats from Campbell Island by the New Zealand government shows that it is possible. On rat-infested islands, the introduction of stoats, weasels, and ferrets as biological controls did not work, and those animals themselves caused havoc. The arrival of cats, either as feral ex-pets or those purposely introduced as rat killers, was also catastrophic for many island birds. The accidental arrival of snakes on the island of Guam seriously depleted several bird populations and killed off the island's flightless rail, which would now be extinct had it not been for captive-breeding efforts. But the disappearance of many flightless birds cannot be blamed solely on the arrival of predators or upon ecological changes that were the consequences of man's actions. Direct assault by human beings has caused the loss of many species, plus the serious reduction of others, and not just insular ones. Hunting is implicated in the extinction of all the giant flightless birds lost in historic times—the moas, elephant birds, Syrian ostrich, and the Tasmanian and Kangaroo Island emus. Sailors en route to and from the East Indies in the seventeenth century filled their vessels' holds with the carcasses of dodos and solitaires, the large flightless pigeons of the Mascarene Islands (Mauritius, Reunion, and Rodrigues) in the Indian Ocean, and played the major role in their extermination. It is assumed that the little-known Reunion flightless ibis (*Borbonibis latipes*), met the same fate, for it was last seen in 1773, and the Rodrigues night heron[1] (*Nycticorax megacephalus*), known only from bones and accounts by Leguat in 1708 and Tafforet in 1726, was last seen in the middle of the eighteenth century.

The Polynesians hunted birds on a massive scale on the islands they colonized. The bones of thousands of moas have been found in Maori middens in New Zealand, and there are authentic accounts from the nineteenth century of their organized hunts for the flightless weka and kakapo. They chased wekas with dogs, catching up to 200 in a single hunt and preserving most of the birds in the fat of others, which they rendered. Kakapos were even easier to kill when they congregated in caves during the winter. Fortunately, both species survived the onslaught, but as a result of its reduced numbers, plus subsequent threats, the kakapo is now one of the world's rarest birds. On both North and South Islands the endemic flightless adzebill, a turkey-sized bird with a massive skull and large wide bill, was exterminated sometime after the Maoris' arrival about 1,000 years ago, but it may not have survived into the Historic Period.

Many species of birds have vanished from the Hawaiian Islands since the arrival there of the Polynesian seafarers; and at least twenty of these were flightless and therefore easy prey for hunters, and for the predators that were eventually introduced. The Polynesians also destroyed most of the island's lowland forest for agriculture. The most northerly of the historically extinct flightless birds, the great auk of the North Atlantic Ocean islands and the spectacled cormorant, which lived on islands in the northwestern Pacific Ocean, were also exterminated through hunting, a misnomer indeed for the slaughter of thousands of helpless and trusting birds.

Historically Extinct Species

Elephant Birds (*Aepyornis* species)

The elephant birds were ancient relatives of the ostrich and other living ratites and were the world's largest known birds. They had very stout legs to support their great body weight, vestigial wings, and small heads with large bills. According to the fossil evidence they lived only in Madagascar, and are thought to have been forest-dwellers. There is disagreement over the number of species of elephant birds, but many authorities recognize three species. The largest of these, *Aepyornis maximus*, was 10 feet (3 m) tall and weighed almost 1,000 pounds (454 kg); its eggs weighed about 18 pounds (8 kg), were up to 13 inches (33 cm) in length, and had a capacity of 2 gallon (9 L) or the equivalent of 150 hen's eggs. Many pieces of these huge eggshells have been found, and on rare occasions a whole egg, which formed the largest single cell in the animal kingdom, larger than any dinosaur egg ever found.

Many *Aepyornis* bones have also been found, mainly in peat deposits in coastal regions of the island, and these prove that it was a very heavy-bodied bird, much stockier and heavier than the ostrich or moas, and with sturdy legs to support its large body weight. It was not a runner like the present-day ostrich, but rather a ponderous bird of the forests. The islanders said it was a peaceful bird and while they collected its eggs, there are no records of them killing it for food. It has featured in literature on several occasions. The giant bird called the roc, which attacked Sinbad the Sailor in the *Arabian Nights*, was based on the elephant bird. Marco Polo wrote of the giant birds of Madagascar and H. G. Wells based his short story "Aepyornis Island" on the bird.

The arrival of humans on Madagascar about 2,000 years ago from east Africa, the Malay Peninsula, and the East Indies (now Indonesia) and the colonization of the island is generally accepted as the reason for the elephant bird's downfall. But the specific causes of their demise are less clear. It has always been thought that they were lost through a combination of hunting, drought, and deforestation, and were either gone or in sharp decline when the Europeans discovered the island. The first Europeans did not step onto the island until about 1500, and the first French settlers came in 1642. Sieur Etienne de Flacourt, the first French governor of Madagascar, claimed the elephant bird still existed in the south when he published his book on the island's natural history in 1658, although he apparently did not see them himself. So they may have survived until the end of the seventeenth century. However, although the extinction of the *Aepyornis* and the New Zealand moas has always been blamed on the human settlement of the islands, and specifically the hunting and egg-collecting, a team of British archaeologists in Madagascar, recently studying the relationship between the elephant bird and the Malagasy people, could find no evidence that it was killed for food or hunted to extinction.

Syrian or Arabian Ostrich (*Struthio camelus syriacus*)

This was the northern-most race of the ostrich, which ranged from southern Syria and Sinai eastward through the Arabian Peninsula and Iraq's Euphrates Valley to Iran. Like many other animals it was once common throughout the Middle East, and had been hunted since antiquity. Historian Flavius Josephus records King Herod the Great hunting ostriches, and many other animals, but whereas it was once more than a match for horsemen, the Syrian ostrich could not outpace vehicles and modern firearms, and sport hunting was the major reason for its extinction. It was still common in the Syrian desert in the nineteenth century and possibly also in the Negev Desert in what is now southern Israel, but its rapid decline commenced after the First World War, when the combination of weapons and vehicles capable of following it across the desert allowed heavy hunting, which it could not withstand. An ostrich was exhibited at the Levant Fair in 1929 that was claimed to have been caught south of Beersheba in the Negev, but it was rarely seen after the mid-1930s and the one killed in 1941 in Hasa Province, on the Saudi Arabian mainland opposite Bahrein, was believed to be the last of its kind. However, in February 1966 a dying ostrich was washed down from the hills near Petra, Jordan, by flash floods.

In the present age of animal reintroductions into their former territories, the ostrich has been returned to the Middle East, but of course a replacement had to be found for the extinct native race. As the North African ostrich (*Struthio c. camelus*) is considered the closest relative it was this race that has been introduced to several regions of the Middle East.[2] It is now well established in Jordan's Shaumari Reserve, to where it was introduced in 1979 and where the population now numbers over fifty. It was also introduced into the Mahazat As Sayd Protected Area in western Saudi Arabia in 1994. Ostriches have been returned to the Negev Desert, specifically to Hai Bar (Israeli for wildlife), a 10,000-acre (4,050 ha) fenced reserve in the Arava Valley created by General Avraham Yoffe in 1968 to house animals mentioned in the bible, many of which were extinct by then. Eventual plans for their reintroduction into the real "wild" include possibly the Ramon Reserve, 250,000 protected acres (100,000 ha) in the central Negev, or possibly into the Halutza sand dunes to the north, bordering Gaza, where there are already a number of commercial ostrich farms.

Tasmanian Emu (*Dromaius novaehollandiae diemenensis*)

The Tasmanian emu was also a smaller and darker bird than the emu of the Australian mainland, and lacked its black neck feathers, but unlike the Kangaroo Island emus its differences were not considered sufficient to justify full species rank, so it remains an endemic race of the Australian emu. It was apparently never an abundant bird, and lived only in the grassy areas and open woodland of the eastern and northern parts of the island, not the rugged, heavily forested western side, which is certainly not typical emu habitat anyway.

Prior to European settlement in Tasmania this emu had coexisted with the aborigines, but the arrival of the convict ships and the subsequent colonization of the island proved disastrous for both bird and man. It was hunted by the early settlers, and then with the development of the penal colony at Port Arthur, soldiers were assigned to shoot emus to feed the convicts and their keepers. It was soon in rapid decline, as a result of direct hunting and the bush fires deliberately set to clear land for agriculture. This was also the era of the development of museum collections and many institutions wanted specimens of the newly discovered species. It has also been suggested that it may have been out-competed by or hybridized with introduced mainland emus, but it seems very unlikely that mainland emus would be shipped to Tasmania in that era, when the island had its own population. The British Museum received specimens in 1828. The Tasmanian emu was last seen in the wild in 1865, and captive birds died out by 1875. By a strange coincidence the last Tasmanian aborigine died the following year.

Kangaroo Island Emu (*Dromaius baudinianus*)

Kangaroo Island lies off the coast of South Australia, and despite the agriculturalization of much of the island's 1,170 square miles (3,000 square km), and the 1 million sheep it supports, it still has substantial areas of bushland, with over 25 percent of its area protected in parks and reserves; and it has no introduced rabbits or foxes. Its emu became isolated from the Australian mainland at the end of the Ice Age about 12,000 years ago, when sea levels rose as the polar ice cap melted. It was discovered in 1802, possibly by Mathew Flinders, who was in the region first, but definitely by French explorer Captain Nicolas Baudin, who arrived soon afterward. It was Baudin who returned to France with a great collection of plants and animals, including emus, and it was scientifically described and named for him. There is a skeleton of the emu in the Paris National Museum of Natural History, but there is little specific information on this bird, except that it was smaller and darker than the mainland emu.

Captain Baudin was a French naval officer and explorer who undertook a "great voyage of scientific exploration" between 1801 and 1803, to explore and chart the southern coast of Terra Australis, as it was known in those days. A conservation park on Kangaroo Island was recently named for him. At the time of his visit he reported that the emu was a plentiful bird on the island.

Francois Peron, a member of the Baudin Expedition, has been described as the first informed zoologist

Tasmanian Emu *Within 65 years of the European colonization of Tasmania the island's indigenous race of the emu was extinct, mainly due to hunting and the loss of its habitat to agriculture. The last wild emus were seen in 1865 and the captive ones died out a few years later. This illustration was made by Louisa Anne Meredith for inclusion in her book* **Tasmanian Friends and Foes,** *published in 1880.*
Photo: Heritage Collections, State Library of Tasmania

to visit Australia. Peron Peninsula and Francois Peron National Park in Western Australia were both named for him. In his report, co-authored with another famous name in biology—wildlife artist Charles Alexander Lesueur—published in 1807, he wrote that the emu was a common bird that lived in the woods and visited the shoreline in the afternoon. Yet W. H. Leigh, in his book *Travels and Adventures in South Australia 1836–38*, published in 1837, said that it had not been seen for ten years, so 1827 is usually quoted as its extermination date, and there were certainly none to be seen when the first settlers arrived in 1836. So this emu completely vanished in a period of about twenty-five years, and its rapid disappearance is one of the major puzzles of animal extinction.

Only sealers and whalers had visited the island regularly since its discovery, and while they hunted the emu it seems unlikely they could have exterminated the bird from such a large island in so few years, despite its habit of visiting the shoreline, where it would have been vulnerable. Bush fires have also been suggested as a contributing factor in its demise. Emus from the mainland—the Australian emu (*Dromaius novaehollandiae*)—were introduced in the 1920s as a replacement.

King Island or Dwarf Emu (*Dromaius ater*)

King Island, with a surface area of 425 square miles (1,100 square km), lies in the Bass Strait between Tasmania and the state of Victoria. A low-lying island with many lagoons, its western coastline is exposed to the roaring forties—the strong winds that circle clockwise around the southern oceans. The King Island emu was quite different from the mainland emu and is considered a distinct species. It was much smaller and lighter, standing only 4 feet, 8 inches (1.4 m) high and weighing about 50 pounds (23 kg), and it had very dark, almost black, feathering. There is little specific information on the natural history of this bird, but once again Captain Nicolas Baudin is responsible for what little is known about it, and he collected the first specimen in 1802. He reported that it was common in open woodland and the grassy areas around lagoons and along the shoreline, where it apparently ate seaweed. At the time much of the western side of the island was heavily forested but this has since been cleared for agriculture. Like Kangaroo Island there are no introduced foxes or rabbits on King Island.

The King Island emu is believed to have been exterminated within a few years of Captain Baudin's visit to the island in 1802, for it was never sighted again, and 1805 is therefore the year normally quoted for its demise. There were extensive bush fires in the early years of the nineteenth century, followed by hunting by visiting sealers and whalers, and if the emus on large Kangaroo Island could not withstand such pressures, the small population on King Island was certainly doomed. There are numerous museum skeletons of this emu and just two skins, one collected by Baudin, which is now in the Paris National Museum of Natural History, and the other in the Turin University Museum.

Moas (*Dinornis* species)

Moas were the giant flightless birds of New Zealand, close relatives of the kiwis and the extinct elephant birds of Madagascar. Their ancestors may have already been flightless when New Zealand broke away from Gondwanaland long ago, but after that time they totally lost their upper arm bones and therefore lacked even vestigial wings. The moas' habits are relatively unknown, but their preserved crops and gizzard contents, complete with stones, prove they were herbivorous like all the living large ratites, and of course New Zealand's limited vertebrate fauna could never have supported so many large birds had they been carnivorous. Browsing and grazing, they took the place on their terrestrial, mammal-free island of the ruminants of other lands. They ranged in size from the 3-foot-tall (90 cm) pygmy moa (*Megalapteryx didinus*) to the giant moa (*Dinornis giganteus*), which stood 10 feet (3 m) high, and was as tall as the largest elephant bird (*Aepyornis*) but not as heavily built, weighing only about 450 pounds (205 kg).

Since their discovery it has been difficult to determine just how many moas there were, and the original estimate of perhaps twenty-seven species has now dwindled to ten species. The uncertainty stemmed from the many different sizes of skeletons that were found, which could have been due to age and to sexual dimorphism—the size difference between male and female—rather than different species. Recent DNA analyses, however, have indeed proved that there was a tremendous difference in size between males and females in the *Dinornis* species, with females double the weight of the males. Consequently, the moas of North Island are now considered to be a single species, *Dinornis novaezealandiae*.

The moas' remains show that they were of two basic types. The most plentiful species were the cursorial or running moas, which were similar in shape to the modern ratites with long, powerful legs, and this group contained both the largest and the smallest species of moas. There was also an intermediate group of massively bodied moas with short, thick legs—and therefore said to be graviportal in habit—which were built for a waddling or rolling gait, rather than for running. It is known from the discovery of pieces of mummified moa skin that their feathers were fine and "hairy" like those of the modern kiwi and cassowary, but the color of their feathers is uncertain due to the possibility of staining or fading during the many years since their death.

No flightless avian carnivores evolved in New Zealand to prey upon the moas, and prior to the arrival of the first Maoris their only predator was the also-now-extinct giant eagle (*Hagapornis moorei*), which was more than equal to the task. It had a wingspan of 10 feet (3 m) and weighed about 32 pounds (14.5 kg). It became extinct also as its prey disappeared, for there was no alternative source of food for such a carnivorous species on the islands.

The exact date of the arrival of the first Polynesians in New Zealand is unknown, but is generally believed to have occurred about 1,000 years ago. The arrival of their "Great Fleet" with a large number of future colonists, occurred in about 1350. The many moa bones and shells found in their middens prove that the Maoris took a

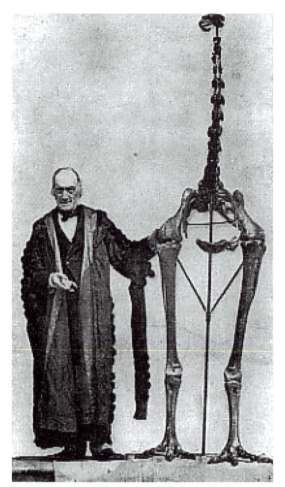

Moa Skeleton   Sir Richard Owen (1804–1892), the famous British paleontologist (who coined the name dinosaur), with a skeleton of the moa Di-nornis giganteus, which stood 10 feet (3 m) tall and was the largest of New Zealand's giant ratites. The moas could not withstand the Maori hunters and most, if not all, were extinct by the end of the eighteenth century.
Photo: Courtesy The Grolier Society

tremendous toll of these birds and their eggs, and it is believed that Maori intertribal warfare did not begin until the scarcity of moas placed more pressure on the fishing grounds and crop-growing areas. Certainly there were few, if any, survivors when the Europeans arrived, and the carbon dating of their bones indicates they were most likely extinct by the beginning of the nineteenth century, with perhaps one of the smaller species surviving in Fiordland into the second half of the century. Sealers described a hunt for very large birds in Fiordland in 1852, although this could have been for the flightless takahe. When Governor Sir George Grey visited Preservation Inlet in 1868, the Maoris there told him of the recent capture of a small moa from a group of half a dozen they had sighted, and several early settlers on South Island also claimed to have eaten moas.

While there is no doubt that the ancestors of today's Maoris were responsible for the extermination of the moas, how long it took them to do this is continually debated. Based upon calculations of the likely population of moas (perhaps 150,000) and their possible rate of reproduction, it has recently been suggested that their extermination would only have taken about 150 years. In any case, when the first European settlers arrived about two centuries ago, there were no moas to be seen; but there were still some very remote and virtually inaccessible regions in the southwest, now called Fiordland, and it was here that some may have survived until perhaps the middle of the nineteenth century.

Atitlan Grebe (*Podilymus gigas*)

The Atitlan grebe is the most recent flightless bird to become extinct, for none have been seen since 1986. It evolved on Lake Atitlan in Guatemala's western highlands, which has been called one of the most beautiful lakes on earth, for the clarity of its water and its dramatic backdrop of volcanic peaks. The grebe was a descendant of the widespread pied-billed grebe (*Podilymnus podiceps*), and was considered by some to be merely a race of that species, hence *Podilymnus podiceps gigas*. However, it was 50 percent larger and quite different in coloration, being mainly a black-and-gray bird with a greenish gloss on its back, whereas the pied-billed grebe is dressed mainly in shades of brown. The small lake, with a surface area of just 50 square miles (130 square km), obviously could not support a large population of grebes, and the birds suffered continual and uncontrolled pressure from a number of sources. Their decline was steady and their eventual loss was foreseeable years before it finally occurred.

The Atitlan grebe was most seriously affected by competition for food from the large-mouth bass (*Micropterus salmoides*), which was released into the lake in 1960, and which was also capable of eating grebe chicks. The upsurge in the bass population coincided with the grebe's decline to just thirty birds in 1983. It was also disturbed by the increasing lakeside human population, increased tourism, entanglement in fishermens' gill nets, and the harvesting of the reed beds where it nested. In addition, as its numbers dropped it may have hybridized with the pied-billed grebe, and its vacant habitat has since been occupied by them. The year of its total disappearence is unknown, as reports of its last sighting vary between 1984 and 1990, and there is even a claim that a survivor was seen in 1996.

Colombian Grebe (*Podiceps andinus*)

This grebe was very similar to the black-necked grebe (*Podiceps nigricollis*) of the Old World, and may even have been its conspecific. It had a chestnut foreneck and upper breast and a dark-golden auricular fan. A former resident of several Andean lakes in the Ubate-Bogota Plateau of eastern Colombia, at elevations between 7,500 feet (2,290 m) and 9,500 feet (2,900 m), this grebe has been restricted since 1960 to Lake Tota, and nineteen specimens were collected during the period 1939–1964. None were seen after 1977, and a survey in 1983 failed to locate any birds. There were several reasons for its extinction, all of them due to the unnatural conditions produced by the development of the lake's immediate surroundings. Lake water was used for irrigation and the resultant drop in water level reduced the grebe's habitat. Its nesting and hiding areas were lost to reed-harvesters, and contamination from soil erosion and chemical pollution from agricultural run-off affected the bird's food supplies, and possibly also its ability to breed. It was also hunted, and the introduction of exotic fish, especially rainbow trout, affected its food supplies.

Spectacled Cormorant (*Phalacrocorax perspicillatus*)

This bird is also known as Pallas's cormorant after the zoologist who described it scientifically, a name which itself superseded the original Steller's cormorant after the German naturalist who actually discovered the bird. George Wilhelm Steller accompanied Vitus Bering on his voyage of discovery to Alaska, and in 1841 was stranded with Bering and the crew of the *St. Peter* after being shipwrecked on an island off the eastern coast of the Kamchatka Peninsula. The island was the largest of the Commander (Komandorski) Islands and has since been named Bering Island. During the long winter months of their stranding they killed the large, black, "almost flightless" waterbirds that lived on the island as they were easy to catch, and in Steller's words "weighed 12–14 pounds, so that one single bird was sufficient for three starving men." Despite this, Vitus Bering and several crew members died during their long winter on the island. The region's great commercial value resulted in an influx of whalers, sealers, and traders seeking the valuable sea otter pelts, plus Aleuts working for the Russian American Company. They killed the cormorants for food and for their feathers, and they were extinct by about 1850. The cormorant had black plumage, a crest, and a pale eye circle, hence its name "spectacled or *perspicillatus.*" Otherwise, nothing is known of the bird, not its biology, exact range, or possible numbers, for Steller was the only naturalist to actually see the cormorant, under circumstances that were hardly conducive to scientific study. Others who saw them later had just one purpose in mind. They are also believed to have lived on several islands in the extreme west of the Aleutian Chain and possibly on the coast of the Kamchatka Peninsula.

The few specimens in the world's museums are there only because of the efforts of Governor Ivan Kupreianov of the Sitka District of Russian Alaska, who acquired seven specimens of the cormorant between 1840 and 1850. When Leonhard Stejneger, curator of the U.S. National Museum, visited the region in 1882, the Aleuts told him they had not seen the cormorant for thirty years.

Auckland Island Merganser (*Mergus australis*)

It is unclear whether this bird had considerably compromised flight or was flightless. It was first seen in 1840, a fish-eating duck with long, thin serated bill typical of the mergansers, that lived in the inland streams and sheltered bays of the Auckland Islands. Since then subfossil remains of the same species have been found on New Zealand's South Island and Stewart Island. On the Auckland Islands its demise has been attributed to the Maori hunting for food, and to predation by introduced cats, dogs, rats, and pigs. Twenty-six specimens of this bird were collected, the last two in 1902 for the British Museum, and it has not been seen since.

Great Auk (*Pinguinus impennis*)

The great auk was the largest member of the family *Alcidae*, a relative of the living puffins and auklets. It nested on several islands in the North Atlantic Ocean, including the Outer Hebrides, Iceland, and Newfoundland, but it swam to warmer seas for the winter, as far south as Florida and Spain. It was the original "pinguin," known to northerners long before the penguins of the southern oceans were discovered.

The great auk's environment encouraged flightlessness, due to the lack of predation in the water and on the isolated islands on which it nested. Loss of flight favored increased size for heat conservation in the cold water, and smaller wings. It was a large black-and-white bird, which stood about 30 inches (76 cm) high and weighed 11 pounds (5 kg); it resembled a penguin, and like them and the other

Great Auk *The demise of the great auk began early in the six-teenth century. For the next 200 years they were slaughtered for their meat, oil, and feathers, and their eggs were collected by the thousands. The last known pair was killed on Eldey Island off Ice-land in 1844, but in 1971 the species finally returned to Iceland in the form of the mounted bird (above) and its egg, which was pur-chased at Sotheby's by the Icelandic Institute of Natural History.*
Photo: Courtesy Pall Stefansson

auks used its shortened wings for propulsion underwater. One of the largest concentrations was on Funk Island off Newfoundland, so named because of the smell from the 200,000 nesting birds.

For almost 200 years, from the beginning of the sixteenth century, these helpless birds were slaughtered wherever sailors had access to their colonies, for their meat, oil, and feathers, and often their chicks and eggs were destroyed just for the fun of it. One sea captain recorded taking 100,000 eggs in a single day, and the adult birds were herded into ships "on sails spread wide as gangplanks." They were salted in barrels or rendered for their oil, others of their kind being used to fuel the fires, and they were killed for fishing bait and for their feathers, which were used to stuff mattresses. The last known pair was killed in 1844 on Eldey Island off the coast of Iceland. In 1971 a specimen finally returned to its former homeland when a mounted great auk, which was killed in Iceland about 1821, was purchased for £9,000 at Sotheby's Auction House by Iceland's Institute of Natural History.

Dieffenbach's Rail (*Gallirallus dieffenbachi*)

This is the other rail of the Chatham Archipelago, endemic to the same three islands—Chatham, Mangere, and Pitt—as the modest rail. The Chatham Islands were home to the Moriori people, believed to have migrated there long ago from South Island, New Zealand. Disease and attacks by the South Island Maori depleted their numbers and the last Moriori died early in the twentieth century.

Dieffenbach's rail has also been considered a race of the banded rail (hence *G. phillipensis dieffenbachi*), from which it is descended, rather than a full species. It was scarce on Chatham Island in 1840, according to E. Dieffenbach who discovered it then, and who also reported that it was preyed upon by the introduced cats and dogs, and hunted by the natives. A medium-sized rail about 11 inches (28 cm) long, it was distinctly colored, with a rufus head with a gray eye stripe and gray throat, and underparts barred with black, white, and buff. It is believed to have contributed to the demise of the smaller Chatham Island rail (*Gallirallus modestus*), which colonized the island first but could not compete with the larger and more dominant newcomer. But within a few years of Dieffenbach's report his rail had also disappeared, after the changes produced by settlement, especially the introduction of cats and the burning of the island's bush and tussock grass to provide sheep pasture. It was last seen alive on Chatham Island in 1872. The date of its extermination on Pitt Island is unknown, and on Mangere Island its disappearance is unrecorded but presumably coincided with the extinction of the Chatham Island rail there at the end of the nineteenth century. There are just two specimens of this rail in museums—at Tring and Bremen—plus numerous subfossil bones.

Chatham Island or Modest Rail (*Gallirallus (Cabalus) modestus*)

Appropriately named, the modest rail was a pale-brown bird with a barred breast, a small rail about 8 inches (20 cm) long. Although sexual dimorphism is rare in rails, in this species the female was more strongly barred and had a much shorter bill than the male. It originally lived on three islands in the Chatham Archipelago—Chatham Island (515 square miles/1,300 square km), Pitt Island (35 square miles/90 square km), and Mangere Island with a land surface of almost 250 acres (100 ha). On the two larger islands it is known only from the bones that have been found there, and it is claimed that it could not compete with the larger and more aggressive Dieffenbach's rail (*Gallirallus dieffenbachi*), yet the two species were apparently compatible on Mangere.

When discovered in 1871, the modest rail survived only on Mangere Island, but this was the era in which the world's museums were clamoring for specimens, and upon discovering this small population of a new bird, twenty-six were immediately shot. But perhaps this was one of the few occasions when such actions were justified, as this was many years before captive breeding to save endangered species was fashionable, and without the museum specimens there would be no visual record of the species, for it was extinct by 1901.

The small island soon lost its vegetation to bush fires and to the introduced rabbits and goats. Cats and dogs then competed for the last rails, together with more museum collectors, and by 1900 practically all of the island's bush and tussock grass was cleared to create sheep pasture.

Wake Island Rail (*Gallirallus wakensis*)

Wake Island, isolated in the western Pacific Ocean between Midway Island and Guam, is a low-lying coral atoll, consisting of three islets. Wake is the largest one, about 5 miles (8 km) long by 3 miles (4.8 km) wide. Its vegetation is open pandanus scrub with grassy clearings and many areas covered with the morning glory vine. Rising no higher than 20 feet (6 m) above sea level, it is a mid-ocean haven for seabirds, and the rail was the only native land bird on the island, which was occupied by Japanese forces from 1942 to 1945 during the War in the Pacific.

The island's native rail was a descendent of immigrant Philippine banded rails (*Gallirallus philippensis*), which found their own way there long ago. It was a small ash-brown bird, about 10 inches (25 cm) long, with a white throat and chin, and narrow white bars on the breast, abdomen, and flanks. Its wings were only 4 inches long and very soft feathered, and it was definitely flightless. The Wake Island rail was a common bird of the shorelines and adjoining patches of beach grass, a very inquisitive bird that was apparently unafraid of humans as it went about its daily routine of digging into the soil in search of invertebrates. It was apparently declining prior to World War II due to predation by cats and rats. In 1945, during the last few months of the war, when the Japanese garrison was beleaguered and isolated and their supply lines from Japan were cut by American forces, the trusting rails were easy prey for the starving soldiers, who ate them to the last one.

MacQuarie Island Rail (*Gallirallus phillipensis macquariensis*)

This attractive bird, with rufus head, white chin and throat, and white eye stripe, was a race of the widespread banded rail (*G. philippensis*) blown to the island years ago. Very similar to its ancestor in New Zealand, about 12 inches (30 cm) long, it was distinguished by its shorter bill, darker wings, stouter legs, and rufus chest band.

MacQuarie Island is the exposed crest of an undersea ridge, the most southwesterly of New Zealand's sub-Antarctic islands, a windswept 60 square miles (155 square km) of tussock grass in the Southern Ocean almost midway between South

Island and the Antarctic Circle. It is a very special place, a Tasmanian State Reserve and a World Heritage Site, with no permanent residents other than the staff of the Australian Antarctic Division's station.

Its rail was reported to be a common bird by the crew of a sealing vessel that visited there in 1879, and it was seen in 1880 shortly after the arrival of cats. Visitors to the island fourteen years later saw no rails, but many rats. The larger weka (*Gallirallus australis scotti*) was also introduced onto MacQuarie Island and thrived because it could defend itself against the rats, but may also have been a competitor of the native rail.

Tahiti Red-billed Rail (*Gallirallus ecaudatus*)

An endemic flightless rail of the islands of French Polynesia in the southeastern Pacific Ocean, this bird is known only from a single painting made by Captain Cook's illustrator, G. R. Forster. There are no known specimens or bones, and Forster's illustration depicts a strikingly marked rail, assumed to be about 10 inches (25 cm) in length. It had grayish-white upperparts, with a black head and face and a white stripe over the eye; its underparts were white with a narrow black band across the chest, and its back and rump were black, spotted with white. It had a heavy red bill and frontal casque rising between the eyes, and orange legs and feet. Rats and cats are believed to have been responsible for its extermination on Tahiti, where it was last seen in 1768, but it survived until the end of the nineteenth century on the small island of Mehetia, to the southeast, apparently because cats were absent there.

Ascension Island Rail (*Atlantisea elpenor*)

A medium-sized rail about 10 inches (25 cm) long, this bird was seen, first and last, by one Peter Mundy, who, with his companions caught and ate six of them in 1656, reporting they "had the taste of roast pig." It apparently succumbed soon afterward to introduced predators. Its flying ancestor was also the ancestor of one of the extinct flightless rails on St. Helena and of the extant Inaccessible Island rail. Ascension Island is a rocky, volcanic peak lying in the middle of the South Atlantic Ocean, midway between Africa and Brazil, 750 miles (1,200 km) north of St. Helena, of which it is a dependency. It has an area of 35 square miles (90 square km) and is mostly a very barren place, of lava flows, ash, and cinder with no standing water, although its highest point—Green Mountain—has lush vegetation. The rails apparently lived along the shoreline where they survived on seabirds' regurgitated food and eggs, and in the desert-like interior, where they ate mainly the eggs and chicks of nesting sooty terns.

The Ascension Island rail is known only from bones and a description and sketch produced by Mundy, a British navigator from Penrhyn, Cornwall, who called at Ascension Island on his way back from India. He described a "strange kind of bird, gray or mottled black and white in color, with red eyes and very imperfect wings, with which they cannot take off of the ground." His rough sketch certainly depicts a rail-like bird. Rats arrived on the island in the late eighteenth century, followed by cats in 1815, and the rails were never seen again.

Tristan Gallinule (*Gallinula n. nesiotis*)

This gallinule or moorhen lived on Tristan da Cunha, the only inhabited island in the Tristan–Gough group in the middle of the South Atlantic Ocean, 1,250 miles (2,000 km) from St. Helena and 1,700 miles (2,735 km) from Capetown. Roughly circular in shape, Tristan has a land area of 48 square miles (125 square km), and is considered the world's remotest island. It is volcanic and very active, the last major eruption occurring in 1961. Gullies and narrow valleys radiate down to the sea from its dominating central peak, which rises to 6,000 feet (1,830 m). The coastline is mostly of sheer basalt cliffs rising up to 1,500 feet (460 m) from the sea, and landing at its small boat harbor can be difficult when seas are rough. The only settlement on the island is Edinburgh, with a population of 300, which lies on the island's only small area of flat land in the northwest. Although discovered by a Portuguese navigator who named the island for himself, and the first person to manage a landing much later was an American, the island was annexed by the British in 1816, and remains a British possession.

The Tristan gallinule, also known as the island hen, was described scientifically in 1861. It was a glossy-black bird with long yellow legs and had a bright-red frontal shield over its bill. It was hunted with dogs when fat in the fall, and the self-styled King of Tristan, one Jonathan Lambert, wrote in 1811 of catching hundreds of the flightless birds. Although in 1873 the members of the Challenger scientific expedition failed to find any, they were obviously still there, as the black rats that colonized Tristan in 1882 from the wreck of the *Henry B. Paul* are blamed mostly for the birds' extinction, which was complete by 1890. Cats, dogs, and pigs now live ferally on the island, but the other islands in the group are fortunately still rat-free, and the race of this gallinule on Gough Island, *Gallinula nesiotis comeri*, still survives there. Some of these gallinules were released on Tristan several times in the 1950s, and are now thriving there also, in place of its native species. London's Natural History Museum has the only two known skins of this bird.

Iwo Jima Rail (*Poliolymnus cinerus breviceps*)

The Iwo Jima rail was a race of the white-browed crake (*Poliolymnus cinereus*), which is a widespread bird in Oceania, the Philippines, and Australasia. A small, dark-brown bird, about 6 inches (15 cm) long, it was mottled with black above and white below. It occurred only on the island of Iwo Jima (Sulphur Island)—a household name because of its place in Pacific warfare rather than for its extinct rail—in the Volcano Islands of the Ryukyu Archipelago, strung between Japan and Taiwan. The island is a highly volcanic submarine caldera dominated by the cone of Mt. Suribachi, with hot springs and boiling mud pools, which has experienced many recent eruptions. Occupied and fortified by the Japanese, with three airfields, it was strategically placed for use in the U.S. bombing of Japan, and the Battle for Iwo Jima claimed 20,000 lives. The island's rail had evolved in the absence of terrestrial predators, bats being the only native mammals. Forest clearance for sugar cane cultivation forced the rails to enter settlements during dry weather to drink at water tanks, where they were preyed upon by domesticated cats, but it was the eventual arrival of rats that caused their extinction by 1925.

Laysan Rail (*Porzana palmeri*)

Laysan is a small, low island in the Northwestern or Leeward group of the Hawaiian chain, about 800 miles (1,280 km) west of Hawaii, and has a land area of 1½ square miles (2.5 square km). Its small rail, just 6 inches (15 cm) long, had pale-brown upperparts and wings, its back and mantle mottled with darker brown; it had pale bluish-brown underparts and face, red eyes, yellow beak, and greenish-yellow legs. It was a common bird when first seen on Laysan Island by Russian sailors in 1828, and was said to be fearless and easily caught. It lived on insects and the eggs of small seabirds such as terns and petrels. The Russians also claimed to have seen the same bird on smaller Lisianski Island (380 acres/153 ha), which lies about 115 miles (185 km) west of Laysan. Regrettably, the provenance of this bird will never be known, as no specimens were collected and it is believed to have perished in the 1880s. However, even assuming it shared ancestry with the Laysan rail, it is unlikely to have been "the same," becoming flightless as it did on a distant island, although the differences may have only warranted subspecific status.

On Laysan Island the rail survived its sporadic settlement by guano diggers, but rabbits were introduced as a source of food in 1903 and within twenty years had destroyed all the island's vegetation. While rabbits were denuding Laysan some rails were captured and "reintroduced" to Lisianski, and to the islands of Pearl and Hermes and islets in Midway Atoll, 375 miles (600 km) west of Laysan, but they survived their relocation only on Midway.

The rail was extinct on Laysan Island by 1924, but survived on Sand Islet (800 acres/225 ha) and Eastern Islet (550 acres/222 ha) in Midway Atoll. Unfortunately, although the rabbits were eradicated on Laysan Island soon after the extinction of the rails, apparently no one thought of returning rails to their former island from the Midway Atoll, and in 1943 rats, which came ashore there from a beached naval landing craft, exterminated the rails there within a year. The last Laysan rail was seen in June 1944 on Eastern Islet.

Hawaiian Rail (*Porzana sandwichensis*)

Also known as the Sandwich rail or crake, and locally as the moho, at only 5½ inches (14 cm) long, it was one of the smallest flightless birds, a pale reddish-brown rail with an olive-brown head and wings mottled with dark brown. It was first collected by members of Captain Cook's Third Voyage in 1779, when it was drawn by William Ellis, the ship's surgeon, and named the Sandwich rail after the original name of the Hawaiian Archipelago. It was thought originally to be one of two species on the islands, but the darker birds seen later proved to be adults of the same species; the first birds collected were all juveniles. The Hawaiian rail frequented the grassy hillsides just below the dense forest on the eastern side of Hawaii and possibly also occurred on Molokai; they had the unusual habit of taking refuge in the holes of native rats when threatened, according to R.C.L. Perkins, an authority on Hawaiian birdlife. Its ancestry and origins are unclear, but it seems most likely to have reached Hawaii from Asia via the islands of Micronesia. The arrival of the Polynesians and dogs, pigs, and rats struck the first blow to Hawaii's flightless birds, considerably reducing their numbers, and the mongooses introduced in 1883 to control rats in the canefields completed the attack. The Hawaiian rail was the last of the islands' flightless species, last collected scientifically in 1864, and last seen alive in 1884.

Kittlitz's Rail or Ponape Crake (*Porzana monasa*)

A small black bird, just 7 inches (18 cm) long, with white-spotted brown wings, red feet, legs, and eyes, this flightless rail lived only on the islands of Kusaie (now Kosrae) and Ponape (now Pohnpei) in Micronesia's Caroline Islands—a large archipelago of small, hot, and humid islands in the western Pacific Ocean. Pohnpei is one of the wettest places on earth with an annual rainfall of 25 feet (7.6 m) and its rails lived in swamps, marshes, and forest undergrowth. Rats that came ashore from whalers in the 1830s, and the loss of traditional beliefs that the rail was sacred (resulting from Christian missionary teachings), are believed responsible for its extinction within twenty years. All that remain of this species are the two specimens collected by F. H. von Kittlitz in 1828, now in the Leningrad Museum.

Mauritian Red Rail (*Aphanapteryx bonasia*)

This rail was a contemporary of the dodo on the tropical Indian Ocean island of Mauritius, to the east of Madagascar. Travelers' accounts, descriptions, and numerous bones in musuem collections are all that remain of this totally flightless, long-legged rail, which more closely resembled an ibis with a long, slightly downward-curving bill, red plumage, and dark-brown legs. A firsthand account came from Jean Francois Cauche, who sailed from Dieppe for Mauritius and Madagascar in 1638 and on his return spoke of "red hens with the beak of a woodcock," and from Johann Hoffman, who lived on Mauritius from 1673 to 1675, and reported "red birds, as big as a common fowle, which cannot fly." Both said the bird was so inquisitive it could be attracted by shaking a red cloth, and when within reach was struck with a stick held in the other hand. They soon discovered, however, that it was preferable not to kill a bird outright as its cries attracted others to the scene. Not surprisingly, it was soon extinct, the "official" date being 1675.

Leguats or Rodrigues Rail (*Aphanapteryx leguati*)

Rodrigues is a small, mountainous island of volcanic origin in the Indian Ocean 400 miles (640 km) east of Mauritius. It is the smallest island of the Mascarenes, and the largest island under the dependency of Mauritius. Only 42 square miles (108 square km) in extent, it is surrounded by a coral reef and was once lushly forested, but most has been cleared for settlement and agriculture in the course of which many native plants were lost, and only two of its endemic birds survive. In 1691, Rodrigues was colonized by a number of French Huguenots led by Francois Leguat, who fled persecution in France. Leguat reported that the rails "have little wingtips without feathers, which renders them unable to fly." They were grayish-white birds with long, straight red bills, red feet and legs, and a red ring around the eye, and were said to be plump all year and very good eating. In addition, their orange-colored fat was believed to cure a number of ailments. Like the Mauritian red rail, this species was also attracted by red objects held in the hand and was therefore very easy to kill. Leguat lived on the island for two years, and also wrote the first account of the Rodrigues solitaire (see page 188). His rail was last seen in 1730.

White Gallinule (*Porphyrio albus*)

The white gallinule is known from just two skins and three accounts and illustrations. It was a striking white bird, with large head, long legs, and short toes, and a red bill, that lived only on Lord Howe Island off the east coast of Australia. Little is known of this unfortunate bird but it is believed to have lived in woodland and must have been flightless, as it was reported to be easily killed with sticks. Lord Howe Island was not discovered by the colonizing Polynesians, nor any Europeans until the French explorer La Perouse reported seeing large congregations of seabirds. The area was investigated by ships of the First Fleet—the vessels that carried convicts from England to Australia under the command of Captain Arthur Philip—and was located by HMS *Supply* in March 1788. Arthur Bowes Smith, surgeon on the convict transport *Lady Penrhyn*, wrote of "fowles or coots, some white, some blue and white others all blue, with large red bills." The blue coots he mentions are undoubtedly purple swamphens, which still live on the island and were the ancestors of the white gallinules, and the blue-and-white birds may have been hybrids between the two. The island then became a port of call for ships destined for Norfolk Island, and the tame and confiding gallinules were slaughtered for food; they were also taken by visiting whalers and sealers, and are believed to have been exterminated before the first settlers arrived in 1834. Lord Howe Island's other, smaller, flightless rail survived on the island until recently, only avoiding extinction through a captive-breeding program.

Assumption Island Rail (*Canirallus cuvieri abbotti*)

This rail, also known confusingly as the white-throated rail, lived only on Assumption Island, a small, now barren island 16 miles (26 km) south of Aldabra in the Indian Ocean. It was a race of the white-throated rail (*Canirallus cuvieri*) of Madagascar that still flies, and which also gave rise to the still-surviving flightless rail of the Aldabra Islands. Assumption Island's original thick vegetation flourished on guano from the large colonies of Abbott's boobies, whose only other known nesting site is on Christmas Island, south of Indonesia. The vegetation was completely cleared to gain access to the rich guano deposits and the rail was last seen early in the twentieth century.

St. Helena Rails

The two extinct flightless rails of St. Helena may, or may not, have survived long enough to qualify as historically extinct species. *Atlantisea podarcas* was a chicken-sized rail which apparently had large wings, but could not fly. It lived along the shoreline and in the heath brush, and was seen to climb and flutter up the steep rock walls, aided by its long claws. It disappeared soon after the island's discovery in 1502. *Porzana astrictocarpus* was a smaller flightless rail, believed to be a descendant of Baillon's crake (*Porzana pusilla*), a resident of the Old World. It is known only from bones but is believed to have been fairly abundant, at least until just before the arrival of humans on St. Helena.

Barred-wing Rail (*Nesoclopeus poecilopterus*)

One of the largest species of rails, reaching a length of 24 inches (60 cm), the barred-wing rail was a somber bird, with brownish-black upperparts and dark ashy-gray belly and chest. It was restricted to the Fijian Islands of Viti Levu and Ovalau, and was related to Woodford's rail, itself now very rare on some of the Solomon Islands to the west of Fiji, which is in keeping with the rail's easterly colonization of the Pacific Ocean islands. Reports vary on this bird's flying ability, and it may not have been totally flightless. It was hunted with dogs and considerably reduced, and was then believed to have been exterminated by the introduced mongoose. Only a dozen specimens were ever collected, all in the nineteenth century, and it was believed extinct until 1973, when one was seen by D. L. Holyoak while making a bird survey in the Vundiawa district of Viti Levu.

North Island Takahe (*Notornis m. mantelli*)

The North Island takahe is known only from subfossil (partly fossilized) bones, dating prior to 1600, and from one possible sighting late in the nineteenth century. Its demise is believed due to the natural extension of the forest into the high tussock grasslands because of climate change in the Holocene (the last 11,000 years of Earth's history, a relatively warm period since the last Ice Age), which reduced its habitat; and possibly also due to hunting in the early years of the Maoris' settlement in North Island, commencing about 1,000 years ago. It may therefore have become extinct prehistorically, but like the South Island race of the takahe it could have survived in remote areas until perhaps the advent of European colonization two centuries ago. Both it and the South Island takahe, which still survives (see Chapter 3) are descendants of the purple swamphen (*Porphyrio porphyrio*).

Dodo (*Raphus cucullatus*)

The dodo was the largest and most unusual pigeon ever known. Early drawings depicted a very plump bird the size of a turkey, weighing 50 pounds (22 kg) and standing about 3 feet (90 cm) tall. It had tiny wings, a heavy, 9-inch-(23 cm)-long hooked beak, and its body feathers were short and bluish-gray or bluish-brown in color. Together with its short, sturdy yellow legs and a tuft of curly feathers which was depicted arising higher on its rump than the normal position of a bird's tail, it had a rather ungainly and almost comical appearance, and was said to be a very clumsy bird. Its name reflects this, for it is derived from the Portuguese doudo, meaning simpleton. However, it is now thought that the dodo may not actually have been as plump as the repeatedly copied illustrations depict, as earlier drawings (from 1601) discovered recently in a museum in the Netherlands show a slimmer bird, which probably weighed only about 33 pounds (25 kg).

Numerous bones of the dodo exist in the world's museums, but there are no intact specimens, the last one having been destroyed deliberately by burning at Oxford's Ashmolean Museum in 1755. But there can be no question about its flightlessness as its breastbone is considered inadequate to support the muscles needed for flight.

The dodo lived only on the island of Mauritius and several small neighboring islets of the Mascarene Islands in the Indian Ocean. Arab traders undoubtedly visited the islands long before they were discovered by Europeans in the sixteenth century, but never settled there, and when the Europeans arrived the islands were still a paradise for the flightless birds, which had evolved in the absence of mammalian predators. The dodo was a confiding and fearless bird, never having known predators of any kind, and despite both the Dutch and Portuguese agreeing that the flesh was unpalatable, tough and of unpleasant flavor, visiting sailors are reported to have hunted it intensively, bringing back fifty birds at a time to their boats to replenish their food supplies. They also shipped specimens back to Europe and India. The full role that human predation played in the extinction of the dodo will never be known, because cats, pigs, and even monkeys were introduced and egg predation then also became a major factor in its decline. It was long believed extinct in Mauritius by about 1650, and on the offshore islands by 1662, but recent investigations place the probable date of its extinction as 1693.

Analysis of genetic material from the dodo by a team from Oxford University shows that it and the Rodrigues solitaire were closely related, and their nearest relatives are the Nicobar pigeon (*Caloenas nicobarica*) of the Nicobar Islands and South East Asia, and the New Guinea crowned pigeons (*Goura* sp.), the changes having resulted from at least 25 million years of isolation and evolution.

Dodo *The most famous flightless bird and the species that epitomizes bird extinctions, yet no complete specimens of the dodo exist. Also, depicted repeatedly in copied illustrations (like the one above by George Shaw, which dates from 1809) as a plump and rather comical bird, earlier drawings recently discovered show a much slimmer bird.*

Photo: General Research Division, New York Public Library, Astor, Lenox and Tilden Foundations

Reunion White Dodo (*Victoriornis imperialis*)

A painting by the Flemish artist Roelandt Savery (1576–1639) portrayed a white dodo that supposedly came from Reunion, a volcanic island in the Indian Ocean about 100 miles (1,600 km) southwest of Mauritius. He may have simply made a mistake, although he had painted normal brown dodos several times so he was familiar with the bird or at least with copying earlier illustrations; yet some of his paintings also depicted dodos with webbed feet, which they certainly did not have.

This bird became known as the Reunion white dodo over the centuries, and was accepted as a separate species, and even given a scientific name. There were also several eyewitness reports of white dodos on Reunion, but the general concensus is that they were influenced by Savery's painting, and following an apparently common practice of the times were getting their islands mixed up. Later artists also painted white dodos, but whether from life or from copying Savery's painting is unknown.

Whereas Savery's dodo could indeed have been from Reunion, it is now considered more likely that it was a white phase or albino dodo, taken there by sailors from Mauritius, or perhaps the bird's color was just a case of artistic license. However, whereas the existence of the dodo and the Rodrigues solitaire have been confirmed by the discovery of many bones and even complete skeletons, there is absolutely no such proof to substantiate the existence of the white dodo as a species in its own right.

Rodrigues Solitaire (*Pezophaps solitarius*)

The Huegenot Francois Leguat, after whom a now-extinct flightless rail (see page 184) was named, also recorded the behavior of the solitaires on the Mascarene Island of Rodrigues early in the eighteenth century. There is no doubt about this bird's existence, for there are numerous other historical accounts and illustrations and complete skeletons exist in several of the world's museums, although, like the dodo, there are no original mounted specimens. The illustrations depict a flightless, turkey-sized bird with predominantly brown or off-white plumage, similar to the dodo but with a smaller head and bill, which was yellow. Its tail did not tuft upward like the dodo's. It was said to be a solitary species, hence the name solitaire, which ate invertebrates dug out of the soil with its sturdy legs.

Leguat reported that the solitaire laid a single egg and that the growing chicks formed a creche while the parents were away. This became a dangerous practice when cats were introduced, and their predations and dry season fires are believed to have been responsible for the solitaire's demise. Also, unlike the unpalatable dodo, it was said to be good eating and was therefore heavily hunted for food.

Leguat said that it was a very territorial bird when nesting and would not allow conspecifics near the nest. Males drove off males and females drove away other females, and if an incubating male was approached by a strange female solitaire he called his partner to see her off. Their bones show that they had a large knob on the wing joint, which was probably used as a weapon for striking each other in their disputes. It was still common in 1730, but by 1755 was reported to be rare, and although in 1761 the island's inhabitants claimed that it still survived, it was never seen again.

Reunion Flightless Ibis (*Threskiornis solitarius*)

Reunion is a small tropical volcanic island in the Indian Ocean, one of the three Mascarene Islands, Mauritius and Rodrigues being the others, and like them is an overseas Department of France. It is a very mountainous island, with the steep forested slopes of Piton des Neiges rising to 9,500 feet (2,900 m). Reunion also had a flightless bird of its own (in addition to the probably mythical white dodo), but this bird was for many years the object of misidentification, which has only recently been resolved.

It was originally known as the Reunion solitaire (*Pezophaps solitarius*), simply because the Portuguese, who were the first to see it in 1613, thought it was the same kind of bird as the solitaire they knew on Rodrigues. The name stuck for many years, and the bird was assumed to resemble the Rodrigues solitaire, even though the French explorer Du Bois, in a book published in 1674, included an illustration of a white bird from Reunion that resembles an ibis rather than a dodo or the other solitaire. Also, unlike those species, which are quite well represented in museums, no remains of the Reunion solitaire have ever been found. Finally, to confuse matters even more, accounts of the solitaire and the probably nonexistent white dodo have often been mingled over the years.

The matter now appears to have been solved by the discovery on Reunion of many remains of an extinct ibis that resembles the living members of the genus *Threskiornis,* but was flightless. So the Reunion solitaire is now considered an ibis and has been renamed the Reunion flightless ibis (*Threskiornis solitarius*), its ancestor probably being the sacred ibis (*Threskiornis aethiopicus*) of Africa. It is believed to have been extinct since early in the eighteenth century.

Broad-billed Mauritian or Blue Parrot (*Lophopsittacus mauritianus*)

Only descriptions and one good drawing of this bird, in Wolphart Harmanzoon's journal of his visit to Mauritius in 1601–1602, plus numerous bones, have survived the passage of time. He depicted a large, heavy-bodied dark bluish-gray parrot, with a long tail, an upstanding crest, and an enormous black bill. Its length was possibly 30 inches (76 cm), so it would have been just a little larger than the living black palm cockatoo (*Probosciger niger*). It is believed to have been restricted to Mauritius, and was probably extinct by 1638.

Mauritius, an island formed by volcanic action in the Mascarene Group in the western Indian Ocean, lies 1,250 miles (2,010 km) from the east coast of Africa. Although initially seen by Arab traders, and later visited by Portuguese navigators, it was first settled by the Dutch in 1598. After razing the island's ebony forests, introducing Javan deer (*Cervus timorensis*), and contributing to the extermination of the dodo, they left in 1710. The island's parrot was monotypic—the only species in its genus—and the bones that were found later include a very large lower mandibular bone, confirming the size of the bird's bill as depicted by Harmanzoon. However, the loose arrangement of the bone tissue and the wideness of the beak suggest it was very weak, and was therefore very likely a fruit eater. There is disagreement over its flying ability, but its bones show that it had a reduced keel on the sternum, and very short wings for a bird of its size, both indications that it was likely flightless.

Table 9.1
Historically* Extinct Flightless Birds
Considerably modified after Greenway 1967, King 1979, and Orenstein 1985

Family	Species & Endemic Insular Races	Distribution	Extinct
Aepyornithidae	Elephant Birds, *Aepyornis* species	Madagascar	1670
Struthionidae	Syrian Ostrich (*Struthio camelus syriacus*)	Middle East	1941
Dromaiidae	Kangaroo Island Emu (*Dromaius baudinianus*)	Kangaroo Island	1827
	Tasmanian Emu (*Dromaius novaehollandiae diemenensis*)	Tasmania	1875
	King Island or Dwarf Emu (*Dromaius ater*)	King Island	1805
Dinornithidae	Moas, *Dinornis* species	New Zealand	(a)
Podicipedidae	Atitlan Grebe (*Podilymnas gigas*)	Guatemala	1990
	Colombian Grebe (*Podiceps andinus*)	Colombia	1977
Phalacrocoracidae	Spectacled Cormorant (*Phalacrocorax perspicillatus*)	Bering Island	1850
Anatidae	Auckland Island Merganser (*Mergus australis*)	Auckland Islands	1902 (b)
Rallidae	Dieffenbach's Rail (*Gallirallus dieffenbachi*)	Chatham Islands	1872 (c)
	Chatham Island Rail (*Gallirallus modestus*)	Chatham Islands	1901
	Wake Island Rail (*Gallirallus wakensis*)	Wake Island	1945
	MacQuarie Rail (*Gallirallus phillipensis macquariensis*)	MacQuarie Island	1890
	Tahiti Red-billed Rail (*Gallirallus ecaudatus*)	Society Islands	1900
	Ascension Island Rail (*Atlantisea elpenor*)	Ascension Island	1656
	Tristan Gallinule (*Gallinula n. nesiotis*)	Tristan da Cunha	1890
	Iwo Jima Rail (*Poliolymnas cinereus breviceps*)	Iwo Jima	1925
	Laysan Rail (*Porzana palmeri*)	Laysan Island	1944
	Hawaiian Rail (*Porzana sanwichensis*)	Hawaii	1884
	Kittlitz's Rail (*Porzana monasa*)	Caroline Islands	1828
	Mauritian Red Rail (*Aphanapteryx bonasa*)	Mauritius	1675
	Leguat's Rail (*Aphanapteryx leguati*)	Rodrigues	1730
	White Gallinule (*Porphyrio albus*)	Lord Howe Island	1834
	Assumption Island Rail (*Canirallus cuvieri abbotti*)	Assumption Island	1910
	Atlantisea podarcas and *Porzana astrictocarpus*	St. Helena	(d)
	Barred-wing Rail (*Nesoclopeus poecilopterus*)	Fiji Islands	1900
	North Island Takahe (*Nortornis m. mantelli*)	North Island	(e)
Alcidae	Great Auk (*Pinguinus impennis*)	N. Atlantic Ocean	1844
Raphidae	Dodo (*Raphus cucullatus*)	Mauritius	1693
	Rodrigues Solitaire (*Pezophaps solitaria*)	Rodrigues	1761
Threskiornithidae	Reunion Flightless Ibis (*Threskiornis solitarius*)	Reunion	1710

Table 9.1
(*continued*)

Family	Species & Endemic Insular Races	Distribution	Extinct
Psittacidae	Broad-billed Parrot (*Lophosittacus mauritianus*)	Mauritius	1638
Acanthisittidae	Stephen Island Wren (*Acanthisitta lyalli*)	Stephen Island	1894

*Extinct in the Historic Period, since 1600
(a) It is unknown how many moa species survived into the Historic Period, but it is likely that they were all extinct by the end of the eighteenth century, except for one of the smaller species, which may have survived in Fiordland until the mid-1800s.
(b) This merganser is believed to have been flightless.
(c) Dieffenbach's Rail was extinct on Chatham Island by 1872, but is believed to have survived on Mangere Island (also in the Chatham Group) until 1901.
(d) Both St. Helena's rails disappeared after the island's discovery in 1502, and possibly before 1600.
(e) The North Island Takahe is known only from subfossil bones, and may have been extinct before 1600.

Stephen Island Wren (*Xenicus lyalli*)

Stephen Island is a rocky and shrub-covered island, with a surface area of just 640 acres (260 ha). It is the most distant island off the northern coast of South Island in New Zealand's Marlborough Sound. At the end of the nineteenth century its only permanent resident was lighthouse keeper David Lyall, but it was also the only known home of a tiny bird which has since been called the Stephen Island wren, a member of the ancient bush wren family of tiny, almost tail-less birds endemic to New Zealand, with no close ties to other species. It apparently skulked among the rocks and was assumed to be insectivorous, yet it had a stout bill.

Lyall was the only person to ever see a living Stephen Island wren, on two occasions in 1894, but he never saw them fly. The same year he also received seventeen dead ones from his cat Tibbles, who caught them at night and deposited them on his doorstep. These tiny brown birds were just 4 inches (10 cm) long and had a greenish-yellow belly and feathers edged with brownish-gray, giving them a mottled appearance. Professional bird-skin collectors avidly acquired all the birds the cat brought in, and their investigation by the Hon. Walter Rothschild showed that they did indeed have all the ingredients of flightlessness—a poorly developed keel, short and rounded wings, and very soft feathers. Not only were they one of the smallest of the songbirds (*Passeriformes*); they were the only known flightless songbird, and were believed to have the smallest range of any known bird species.

As Lyall did not see them more often, it is thought they may have been nocturnal or at least active in the twilight of dusk and dawn. After the cat's nocturnal hunting, the wren was never seen again, alive or dead. This gave the Stephen Island wren the dubious distinction of being discovered and then exterminated by the time its discovery was made known. Of the birds the cat brought home, ten specimens survive in the world's musuems.

Notes

1. It is uncertain whether this species was just a poor flier or completely flightless.

2. If the recent reclassification of the ostriches by Sibley and Monroe, in which only two races are recognized, is accepted, then the ostriches that occurred throughout North Africa, Syria, and the Arabian Peninsula belonged to the same race anyway.

Glossary

Acarinatae

One of the three major groups of the bird kingdom—the Class *Aves*—and containing the flight-less ratites, in which the keeled sternum (to which the flight muscles were once attached) has degenerated; these birds are flat-chested and therefore have no *"carina,"* Latin for keel. The other subclasses contain the penguins and the flying birds that have a well-developed keel, to which the flight muscles are attached. It has recently been suggested that these subclasses are inaccurate and should be based upon the structure of the palate. See also Paleognatae and Neognatae.

Adaptive radiation

On remote islands the arrival of species is often so infrequent that the original colonizer can evolve to fill vacant niches and in doing so changes in appearance and behavior, eventually becoming new species. The fourteen species of Darwin's finches on the Galapagos Islands are the classic example of adaptive radiation, all changing so much since their ancestors' arrival that their common ancestry was not immediately recognized.

Altiplano

The high-altitude dry plains above the tree-line in Argentina, Bolivia, Chile, and Peru; the home of Darwin's rhea.

Altricial

Chicks that are helpless, hatched with their eyes closed, and usually naked with little or no down feathers. They are incapable of locomotion so cannot leave the nest, and are therefore totally dependent upon the parents for warmth and food. The young of all passerine or perching birds are altricial.

Alula

Also called the bastard wing or false wing, it is a tuft of small feathers arising from the first digit of a bird's wing.

Antarctic Convergence
The region in the Southern Oceans where the cold surface water flowing north from the Antarctic meets the warmer sub-Antarctic waters flowing south, about 1,000 miles (1,610 km) north of the continent.

Apterous
Lacking wings or wing-like extensions, from the Greek *apteros*, meaning wingless. Unfortunately, it has become synonymous with flightlessness in birds, but excluding the wingless kiwis, and the emus and cassowaries which have tiny vestigial wings, all flightless birds have wings of varying size, some quite large, but they are unusable due to the deterioration of their flight muscles.

Archaeopteryx
The feathered, bird-like animal whose fossil was discovered in Bavaria in 1860, and is believed to be 150 million years old. Until recently, it was thought to be the missing link between the reptiles and birds, but it is now generally accepted that birds descended from the dinosaurs.

Artificial incubation
Incubation of birds' eggs by means other than the parent or a foster parent, generally, therefore, in an incubator, with controlled temperature and humidity.

Asymmetry
Lacking symmetry, when parts of the body are unequal. Created artificially in birds by pinioning one wing to produce "lopsidedness" and the inabilty to fly. See Pinioning.

Atrophy
The degeneration of an organ through lack of use. In flightless birds the atrophy of the pectoral muscles, which are used in flight, and the eventual changes to the sternum or breastbone—the flattening of the keel to which the muscles are attached.

Barbs
The numerous slender, closely arranged parallel structures that form the vane on either side of the feather shaft.

Barbules
The projections from the barbs of a flight feather, which interlock with adjacent barbules by means of hooklets, to form the vane of a feather. Down feathers do not interlock and are therefore soft and fluffy for better insulation.

Benguela Current
The current of cool water from the Antarctic that flows around the tip of Africa and northward along the west coast until forced outward by the bulge of West Africa, when it becomes the South Equatorial Current. The rich upwelling of nutrients from the polar seas supports an enormous population of fish, which in turn feeds the once numerous black-footed or jackass penguin.

Bird fancy
The hobby of keeping and breeding birds, generally domesticated ones such as zebra finches, canaries, and fancy poultry, and pigeons. Those who keep and breed wild birds are called aviculturists.

Breed
The domesticated animal equivalent of subspecies in wild animals. For example, the leghorn, Rhode Island red, and Sussex are all breeds of the domesticated chicken. Its ancestor the wild red jungle fowl

(*Gallus gallus*) is represented by several subspecies across its range, such as the Burmese jungle fowl (*G.g. jabouillei*) and the Javan junglefowl (*G.g. bankiva*).

Broken-wing display
Simulating the disability of having a broken wing, a tactic used by many terrestrial birds to draw predators away from their nest or young. Even the ostrich and rhea, both grounded for many years, still use this tactic.

Brood patch
A bare patch that allows birds to have direct contact with their eggs, without feathers preventing heat transmission. Most female birds (and some males that also incubate the eggs) develop a brood patch during the breeding season. Changes in their hormone levels result in the down feathers on the bird's stomach falling out, or loosening and being pulled out. Ducks actually use them to line their nests. Expansion of the blood vessels to the bare patch then increases the transference of heat to the eggs and incubation can proceed.

Campos
The dry, grassy plains studded with trees and shrubs in central-eastern South America.

Carinate
Having a keeled sternum or *carina* (Latin for keel) to which the flight muscles are attached, and therefore being able to fly (except for the penguins, who have retained the keel and well-developed pectoral muscles for powering their wings underwater). Also in birds that have more recently become flightless but so far have lost only their muscles and not their keel.

Carnivore
A member of the order *Carnivora*, which includes the cats, dogs, and hyenas. However, it is also used to denote any animal that eats animal protein such as fish, crustaceans like crabs and krill, and the meat of mammals, reptiles, and amphibians. Carnivorous flightless birds include the fish-eating penguins and cormorants.

Carpals
The bones of the wrist, which are fused in birds (to restrict the motion to one plane) together with the metacarpals to form the carpometacarpus and modified for the attachment of the feathers.

Cerebellum
The division of the brain that controls a bird's posture and balance, maintaining and coordinating muscle movement. It is essential for flight.

CITES
The Convention on International Trade in Endangered Species, which is the world's major conservation law, ratified by most nations of the world since its inception in the early 1970s.

Clavicle
The collarbone, which connects the sternum (breastbone) with the scapula (shoulder blade). In birds it is fused to form the furcula, a large Y-shaped bone that braces the flight stroke, and which is commonly called the wishbone.

Convergence
The development of similar morphological (form and structure) traits in unrelated or very distantly related organisms as a result of adapting to the same way of life. The lobed toes of the unrelated grebes and giant coot as an aid to swimming and diving are an example in flightless birds.

Flightlessness is itself an adaption to a terrestrial way of life, which has resulted in similar changes in many birds, related and unrelated.

Crepuscular
Active at dusk and dawn, or seminocturnal.

Cretaceous
The last period of the Mesozoic Era, from 140 to about 65 million years ago, during which the dinosaurs died out and birds began to evolve.

Crop
A bird's pouch-like enlargement of the lower esophagus for the temporary storage of food.

Cursorial
Adapted or modified for running, with strong legs and well-developed thighs, such as the ratites—ostrich, rhea, cassowary, emu, and kiwi.

DNA
Deoxyribonucleic acid, molecules that carry the genetic information essential for the organization and functioning of the cell, and for controlling the inheritance of characteristics.

Domestication
The results of the continual control of wild animals by man over many generations, producing changes in behavior, morphology, and physiology.

Down feathers
The soft and fluffy feathers that are important for insulation, especially in the nestling bird. Their barbs do not link to form a closed vane and stiffen them as they do in flight feathers.

Eclipse plumage
The body feathers that grow prior to male ducks moulting all their flight feathers. To reduce their visibility and vulnerability during this flightless period they moult their brightly colored breeding plumage and replace it with duller feathers, and so resemble the dull females. When they have regrown their flight feathers they moult again and regrow their normal-colored plumage.

El Niño
A disruption of the oceanic and atmospheric systems in the tropical Pacific Ocean resulting in increases in sea temperature, which affects weather systems around the world. Drought and bush fires in Australia and high rainfall on the normally dry Peruvian coast have been blamed on El Niños. The increase in water temperature on the northwestern coast of South America has affected fish stocks and the resident aquatic animals, including the penguin and flightless cormorant.

Endemic
Native to or confined to a particular region. The kiwi is endemic to New Zealand, the kakapo is endemic to New Zealand's South Island, and the now extinct modest rail was endemic to New Zealand's sub-Antarctic Chatham Islands.

Esophagus
The muscular tube that connects the throat to the stomach.

Ethiopian Region
Africa south of the Sahara, one of the world's major zoogeographic regions.

Euphausids—see Krill

Falkland Current
A branch of the Antarctic Circumpolar Current flowing north along the coast of Argentina, its cold water (at 41°–66° F) providing ideal conditions for several species of penguins on the Falkland Islands and the Magellanic penguin on the mainland coast.

Family
Zoologically, a subgroup of an Order. For example, the Order *Gruiformes* contains bird families such as *Rallidae* (rails), *Rhynochetidae* (the kagu), and *Mesitornithidae* (mesites). There are about 216 bird families.

Feral
A domesticated animal that has returned to the wild and is living free, although not necessarily in its country of origin.

Fledging
To grow the first feathers needed for flight.

Fledgling
A young altricial bird that can leave the nest but is still dependent on parental care.

Flight
The ability to become airborne and to maintain height and speed.

Flight feathers
A bird's flight feathers are its primaries and secondaries, the wing feathers used to gain lift. Tail feathers are also considered flight feathers, but really in name only, for although they have an important role in maneuvering—as a rudder—birds that have lost their tails have no difficulty flying. Flight feathers must maintain their shape to resist air pressure, and the vane—the actual feather on either side of the "stem" or shaft—is composed of barbs that are interlocked by barbules—a very light but strong arrangement.

Flight muscles
The pectoral or breast muscles, attached to the keel and well developed in the flying birds. The large pectoralis muscle pushes the wing down in flight and the smaller supracoracoideus raises the wing for the return stroke, assisted by air pressure. They have degenerated in the flightless birds.

Flightless
The inability to fly, in birds due first to the degeneration of the flight muscles, then the reduction of the keel, followed eventually by some degree of wing reduction, but not necessarily total loss of wings.

Flippers
Ornithologically, the modified wings of the penguins for "underwater flight" coupled with the retention of the keel and powerful pectoral muscles, despite being flightless in the accepted sense.

Glycogen
Glucose converted for storage in the muscles and liver. It is used as fuel for infrequent bursts of activity (of the chicken's white breast meat, for example, when it occasionally flaps its wings, but cannot fly) as opposed to the more sustained activity which its legs receive. The leg muscles are fuelled by myoglobin—a chemical that stores oxygen and allows sustained muscle contraction—the effect being to darken the drumsticks, hence dark meat.

Gondwanaland
The southern land mass that broke up in the middle of the Cretaceous Period, about 100 million years ago, to form India and the continents of Australia, Africa, Antarctica, and South America.

Grazers
Ornithologically, birds that eat grass. Most of the geese are grazers, as are some ducks such as the American widgeon. However, as they cannot digest plant fiber (cellulose and hemicellulose), this is undertaken in their enlarged caeca by symbiotic microflora. This is similar to the manner in which ruminant mammals ferment and digest fiber in their four-chambered stomachs.

Gregarious
Birds that live in or form groups with others of their own kind. Some, such as the rhea, are gregarious for most of the year, but nest singly; whereas the penguins go their own way while out at sea, but congregate in large colonies to nest. See Social.

Hatching synchronization
In some birds, such as the ostrich and rhea, in which the eggs are laid over many days but the chicks must leave the nest soon after hatching for safety and to feed, the embyronic chicks call to each other in the egg and synchronize their hatching, often shortening the incubation period of some eggs by several days.

Historic Period
The last four centuries. A period during which the actual appearance of numerous birds that became extinct then is recorded in paintings, drawings, and eyewitness accounts. With very few exceptions, such as the rock paintings of the extinct moas, bones or fossils are generally all that remain of extinct species from prior to the seventeenth century—the prehistoric period.

Holocene
The post-glacial period. The most recent geologic era, from the last Ice Age about 10,000 years ago to the present day.

Humboldt Current—see Peruvian Current

Incubation period
The hatching period of eggs, calculated from the time of exposure of the brood patch (not necessarily the time of egg laying) when the egg comes into contact with the bird's body unprotected by the insulating feathers, to the emergence of the chick. See Brood patch.

Indigenous
Native to a region, originating where it is found.

Insular
Relating to or situated on an island.

Invertebrates
Animals without backbones.

Jungle fowl
The wild ancestor of the domesticated chicken. A member of the order *Galliformes* from India and Southeast Asia.

Keel
The median ridge on the breastbone of birds that fly, and penguins.

Krill
A Norwegian word for whale food. Krill are shrimp-like marine invertebrates, which are the major component of zoo- or animal-plankton, and which forms the bulk of the diet of many penguins. In southern waters the Antarctic krill (*Euphausia superba*), a large species reaching 2½ inches (6.5 cm) in length, lives in huge congregations feeding on the tiny phyto- or plant-plankton. Their biomass is estimated to be at least 500 million tons; they are regarded as an important human food source and are already being harvested commercially.

Laurasia
Part of the enormous hypothetical land mass known as Pangaea, which separated between 250 and 200 million years ago into Gondwanaland in the south and Laurasia in the north.

Lek
The site or "arena" where male birds of certain promiscuous species display collectively to attract females. Examples—cock of the rock, ruff, and in flightless species the kakapo.

Lift
Getting airborne. Lift is produced as the wing deflects air downward, and the amount deflected and the resultant lift depends upon the surface area of the wing and its shape.

Llanos
The grasslands of northern South America, where the hoatzin occurs in swamps and riverine woodland.

Marsupials
Primitive implacental mammals in which the young are born after a very short gestation period and complete their development in the mother's pouch or marsupium. Examples that have an impact on flightless birds are Tasmanian devils and the native cats.

Megapodes
Birds of the order *Megapodiidae*, in which the eggs are laid in a mound of soil and leaves, in volcanic earth or sand on a sun-warmed beach. They are incubated by the sun's heat, by geothermal activity or bacterial decomposition within the mound. The chicks hatch after a long incubation period (up to 70 days) and are able to fly soon after they dig themselves out.

Metacarpals
The bones of the "hand," which are fused in birds, and from which the primary feathers grow.

Migration
The movement of birds from one country or locality to another, usually long distance, between winter and summer ranges, to traditional moulting sites or between islands in Oceania (except for the semiflightless Henderson Island fruit dove) to take advantage of fruiting trees.

Monogamy
Mating with just one of the opposite sex, the most common mating arrangement in birds.

Monotypic
A taxon with just one immediately subordinate taxon, for example, a family with just one species, such as the kagu.

Moult
The periodic shedding of feathers, generally once annually, but twice in some penguins and

in ducks, which have an eclipse plumage. The word is now accepted to mean feather replacement also.

Muttonbirds
The local name given to the young of two species of burrow-nesting shearwaters—in New Zealand the sooty shearwater and in Australia the short-tailed shearwater—which are harvested annually in the thousands.

Neognatae
A subclass of the bird kingdom *Aves* in which the palate (the upper surface of the mouth, which separates the oral and nasal chambers) is characterized by a small vomer bone and large palatine bones. It includes all flying birds and those that have recently become flightless. See Paleognatae.

Neotropics
A zoogeographic region incorporating tropical Central and South America from Mexico to southern Brazil.

Nidicolous birds
Young birds that remain in the nest for some time after hatching, but need not be naked (psilopaedic) or helpless (altricial).

Nidifugous birds
Young birds that leave the nest soon after hatching. They therefore have varying degrees of independence (precocial) and have feathers or down (ptilopaedic).

Nocturnal
Active during the hours of darkness.

Notogaea
A zoogeographic region encompassing Australia, New Zealand, and the islands of eastern Indonesia.

Oil gland
An oil-secreting gland situated at the base of the tail in most birds. Preening promotes the spreading of oil from the gland onto the feathers, keeping them flexible and waterproof.

Oligocene
An epoch in the middle of the Tertiary Period, from 40 to 25 million years ago.

Order
In biology a taxonomic category between a Class and a Family; for example, Class *Aves* (birds), Order *Gruiformes* (cranes and rails), Family *Rallidae* (rails).

Palate bone
The bony and muscular partition between the oral and nasal cavities—the roof of the mouth.

Paleognatae
A subclass of the bird kingdom *Aves*, containing the ratites and the tinamous. Their palates (the upper surface of the mouth that separates the oral and nasal chambers) are characterized by a large vomer bone and small palatine bones. See Neognatae.

Pectoral
Situated on or pertaining to the chest. In flying birds the well-developed pectoral muscles and the keeled sternum or breast plate known as a keel, both of which have degenerated to varying degrees in flightless birds.

Peruvian Current
The cold water flowing north from Antarctica along the western coast of South America, making it and the Galapagos Islands (situated on the equator) a suitable environment for penguins. Also called the Humboldt Current.

Physiology
The branch of biology dealing with the functioning of organisms.

Pinioning
The permanent prevention of flight through the removal of tendons or amputation of the last segment of one wing.

Pliocene
The fifth epoch of the Cenozoic Era, from about 5.0 to 1.8 million years ago.

Polyandry
Polygamy where a female mates with several males. A rare form of mating occurring in just 1 percent of all birds. The female lays a clutch for each male and then leaves them to incubate the eggs and raise the chicks. She may sometimes have two male partners at once, which is called simultaneous polyandry.

Polygamy
Birds that have multiple mates of the opposite sex. It occurs in about 3 percent of all bird species.

Polygyny
An uncommon form of polygamy in birds where a male mates with two or more females. It occurs in perhaps 2 percent of all birds.

Polytypic
Having several forms, especially subspecies.

Precocial
Young birds that hatch with their eyes open, a covering of down, and even feathers in a few species. They are able to leave the nest soon after hatching, but they vary in their degree of independence. Megapode chicks are totally independent upon hatching. Dependent species require brooding for warmth, but vary in their feeding habits. Ducklings stay with their parents but find their own food, most pheasant chicks initially have to be shown what to eat, and rails feed their chicks for two or three weeks.

Primaries
The major feathers involved in flight. The average flying bird has ten primaries per wing, whereas the ostrich, surprisingly, has sixteen.

Proventriculus
The glandular stomach, between the crop and the ventriculus or gizzard, which stores and digests a bird's food.

Psilopaedic
Young birds that are naked or just have sparse down upon hatching.

Psittacine
Members of the "parrot" family (*Psittacidae*), which includes parrots, macaws, cockatoos, lovebirds, and many others.

Ptilopaedic
Young birds that are covered with down upon hatching.

Pygostyle
The plate of bone that forms the posterior end of the vertebral column in most birds, from which arise the tail feathers.

Race—see Subspecies

Ratites
Birds of the subclass *Acarinatae* (no *carina* or keel), due to their flat, raft-like breasts, unlike the flying birds, which have a sternum shaped like a boat's keel, to which the flight muscles are attached. Ratites are characterized by their large bodies, powerful legs, and the great reduction of their wings and flight muscles.

Roaring forties
The westerly winds that circle the earth between 40° and 50° S latitude (just north of Antarctica) unobstructed by land.

Secondaries
The other important feathers involved in flight (with the primaries), of which most flying birds have ten per wing.

Social
Tending to move or live together in groups or colonies. See Gregarious.

Squabs
The nestlings of pigeons and doves, and the shearwaters harvested commercially as muttonbirds for human consumption.

Subspecies
A subdivision of a species; a population distinguishable from other such populations and capable of interbreeding with them. Synonymous with "race."

Subtropical Convergence
The limit of the cool sub-Antarctic water flowing north, about 1,000 miles (1,610 km) north of the Antarctic Convergence and therefore 2,000 miles (3,220 km) from the polar ice cap.

Symbiotic bacteria
Bacteria that may benefit the host.

Taxon
A group or category at any level in the system of classifying animals, for example, family, genus, species.

Tertiary
The first period of the Cenozoic Era, from 65 to 1.8 million years ago.

Tubenoses
Pelagic birds of the Order *Procellariformes* (albatrosses, fulmars, petrels), in which the nostrils are extended externally in short tubes, which excrete the salt that is unavoidably ingested during their life at sea. Believed to be ancestors of the penguins.

Type species
The species that best exemplifies the characteristics of its genus.

Uropygial gland—see Oil gland

Vane
This is the actual feather on either side of the central shaft or quill, formed by parallel rows of barbs interlocked with barbules and hooklets. When the barbs are separated they can be rejoined by running the feather between the fingers and thumb, or as the bird does with its beak when it preens.

Vascular heat exchange
The means by which Antarctic penguins keep warm in sub-zero temperatures. Arteries carrying blood at body temperature run alongside veins that are returning cooled blood from the extremities, thus warming the returning blood and holding the temperature of their flippers and feet above freezing.

Veldt
The elevated, open grasslands of southern Africa.

Wing-loading
The ratio of the weight of a bird to its wing area.

Wing feathers—see Flight feathers

Zoogeographic Regions
The divisions of the earth characterized by distinct forms of plant and animal life, for example, Ethiopian Region—Africa south of the Sahara. Zoogeography is the study of the geographic distribution of animals, and the causes and effects of such distribution.

Bibliography

Berger, A. J. *Hawaiian Bird Life*. University of Honolulu Press, Honolulu, 1972.

Bird Biology. http://animaldiversity.ummz.umich.edu/site/index.html.

Bond, J. *Birds of the West Indies*. Collins, London, 1960.

Bourne, W.R.P. Penguin breeding failures and mortality. *Oryx* 21 (1986): 112.

Brown, L. and Amadon, D. *Eagles, Hawks and Falcons of the World*. Country Life, Feltham, 1968.

Cabot, J. *Family Tinamidae (Tinamous)*. In *Handbook of Birds of the World*. Vol. I. Edited by J. del Hoyo, A. Elliott, and J. Sargatal. Lynx Editions, Barcelona, 1992.

Campbell, B. and Lack, E. (Eds.). *A Dictionary of Birds*. Buteo, Vermilion, 1985.

Collar, N. J., Crosby, M. J., and Stattersfield, A. J. *Birds to Watch* 2. BirdLife International, Cambridge, 1994.

Cracraft, J. Evolution of the ratites. *Ibis* 116 (1974): 494.

Davis, L. S. and Darby, J. T. (Eds.). *Penguin Biology*. Academic Press, San Diego, 1990.

Day, D. *The Doomsday Book of Animals*. Viking, New York, 1981.

Delacour, J. *The Waterfowl of the World*. Country Life, Feltham, 1961.

Diamond, A. W. *Studies of Mascarene Birds*. Cambridge University Press, Cambridge, 1987.

Dodo. http://Birds.mu/Extinct/Dodo.htm.

Elton, C. S. *The Ecology of Invasions by Animals and Plants*. Methuen, London, 1958.

Falla, R. A., Sibson, R. B., and Turbot, E. G. *A Field Guide to the Birds of New Zealand*. Houghton Mifflin, Boston, 1967.

Feduccia, A. F. Flightless birds. In *Dictionary of Birds*. Edited by B. Campbell and E. Lack. Buteo, Vermilion, 1985.

Forshaw, J. M. *Parrots of the World*. TFH, Neptune, 1977.

Fullagar, P. J. The Woodhens of Lord Howe Island. *Avic. Mag.* 91 (1985): 15–30.

Fuller, E. *Extinct Birds*. Facts on File Publications, New York, 1987.

Furness, R. W. *The Skuas*. Poyser, Calton, 1987.

Greenway, J. C. *Extinct and Vanishing Birds of the World*. Viking, New York, 1967.

Hachisuka, M. *The Dodo and Kindred Birds*. Witherby, London, 1953.

Harris, M. P. *A Field Guide to the Birds of Galapagos*. Collins, London, 1974.

Harrison, P. *Seabirds—An identification guide*. Houghton Mifflin, Boston, 1983.

Hays, C. The Humboldt Penguin in Peru. *Oryx* XVIII (1984): 92–95.

Howard, R. and Moore, A. *A Complete Checklist of the Birds of the World.* Oxford University Press, Oxford, 1980.

Howard, W. E. *Control of Introduced Mammals in New Zealand.* NZ DSIRO Info Ser. #45, 1966.

Kemp, A. C. *The Hornbills.* Oxford University Press, Oxford, 1995.

King, W. B. *Red Data Book.* Vols. I and II. IUCN, Morges, 1979.

King, W. B. *Endangered Birds of the World.* ICBP Red Data Books. Smithsonian Institute Press, Washington, DC, 1981.

Kiwi Recovery. http://kiwirecovery.org.nz.

Kiwis. http://www.kiwi.bird.freeservers.com.

MacDonald, J. D. *Birds of Australia.* Reed, Sydney, 1973.

Merlen, G. The 1982–83 El Niño: Some of its consequences for Galapagos wildlife. *Oryx* XVIII (1994): 210–214.

Moas. http://www.duke.edu/~mrd6/moa/.

Morton, H. V. *In Search of South Africa.* Methuen, London, 1948.

Murphy, R. C. *Oceanic Birds of South America.* American Museum of National History, New York, 1936.

New Zealand's Flightless Birds. http://www.terranature.org/flightlessBirds.

New Zealand's Penguins. http://www.penguins.net.nz/.

Nitecki, M. H. (Ed.). *Normal Extinction of Isolated Populations.* University of Chicago Press, Chicago, 1984.

Oliver, W.R.B. *New Zealand Birds.* Reed, London, 1974.

Olson, S. L. and James, H. F. Fossil birds from the Hawaiian Islands. *Science* 217 (1982): 633–635.

Orenstein, R. I. Extinct birds. In *Dictionary of Birds.* Edited by B. Campbell and E. Lack. Buteo, Vermilion, 1985, 660–668.

Ostrich. http://digimorph.org/specimen/Struthio_camelus/skull/.

Ripley, D. *Rails of the World.* Godine, Boston, 1977.

Roots, C. *Animal Invaders.* David & Charles, Newton Abbott, 1976.

Rowe, B. E. The Kiwi *Apteryx australis*: An extreme in egg size. *International Zoo Yearbook* 20, Zoological Society of London, London, 1980.

Ryan, P. G. and Cooper, J. Rockhopper penguins and other marine life threatened by driftnet fisheries at Tristan da Cunha. *Oryx* 25 (1991): 76–79.

Sibley, C. G. and Monroe, B. L. *Distribution and Taxonomy of Birds of the World.* Yale University Press, New Haven, 1990.

Smythies, B. E. *The Birds of Borneo.* Oliver & Boyd, Edinburgh, 1968.

Soper, M. F. *New Zealand Birds.* Whitcombe & Tombs, Christchurch, 1972.

Sparks, J. and Soper, T. *Penguins.* David & Charles, Newton Abbott, 1987.

Takahe. http://forests.org/archive/spacific/rarebinz.htm.

Todd, F. S. *Waterfowl: Ducks, Geese and Swans of the World.* Sea World Press, San Diego, 1979.

Tuck, G. and Heinzel, H. *A Field Guide to Seabirds of Britain and the World.* Collins, London, 1980.

Turbott, E. G. (Ed.). *Buller's Birds of New Zealand.* Whitcombe & Tombs, Christchurch, 1967.

Williams, G. R. Marooning—A technique for saving threatened species from extinction. *International Zoo Yearbook* 17, Zoological Society of London, London, 1977.

Index

About the Author

CLIVE ROOTS has been a zoo director for many years. He has traveled the world collecting live animals for zoo conservation programs. Roots has acted as a masterplanning and design consultant for numerous zoological gardens and related projects around the world, and has written many books on zoo and natural history subjects.